Rodica Branzei · Dinko Dimitrov · Stef Tijs

Models in Cooperative Game Theory

Second Edition

Springer

Professor Dr. Rodica Branzei
Faculty of Computer Science
Alexandru Ioan Cuza University
16 Berthelot St.
700483 Iasi
Romania
branzeir@infoiasi.ro

Professor Dr. Stef Tijs
CentER and Department of Econometrics
and Operations Research
Tilburg University
5000 LE Tilburg
Netherlands
s.h.tijs@uvt.nl

PD Dr. Dinko Dimitrov
Chair of Microeconomics
Faculty of Law and Economics
University of Bayreuth
95440 Bayreuth
Germany
dinko.dimitrov@uni-bayreuth.de

Originally published as Volume 556 in our Series 'Lecture Notes in Economics and Mathematical Systems'

ISBN 978-3-540-77953-7 e-ISBN 978-3-540-77954-4

DOI 10.1007/978-3-540-77954-4

Library of Congress Control Number: 2008921947

Production: le-tex Jelonek, Schmidt & Vöckler GbR, Leipzig
Cover-design: WMX Design GmbH, Heidelberg

Printed on acid-free paper

9 8 7 6 5 4 3 2 1

springer.com

Preface

Cooperative game theory is a booming research area with many new developments in the last few years. So, our main purpose when preparing the second edition was to incorporate as much of these new developments as possible without changing the structure of the book. First, this offered us the opportunity to enhance and expand the treatment of traditional cooperative games, called here crisp games, and, especially, that of multi-choice games, in the idea to make the three parts of the monograph more balanced. Second, we have used the opportunity of a second edition to update and enlarge the list of references regarding the three models of cooperative games. Finally, we have benefited from this opportunity by removing typos and a few less important results from the first edition of the book, and by slightly polishing the English style and the punctuation, for the sake of consistency along the monograph. The main changes are:

(1) Chapter 3 contains an additional section, Section 3.3, on the average lexicographic value, which is a recent one-point solution concept defined on the class of balanced crisp games.

(2) Chapter 4 is new. It offers a brief overview on solution concepts for crisp games from the point of view of egalitarian criteria, and presents in Section 4.2 a recent set-valued solution concept based on egalitarian considerations, namely the equal split-off set.

(3) Chapter 5 is basically an enlarged version of Chapter 4 of the first edition because Section 5.4 dealing with the relation between convex games and clan games with crisp coalitions is new. Additionally, the structure of Section 4.2 of the first edition has been changed for the sake of consistency with the other two models, resulting in the actual Section 5.2.

(4) Chapter 7 contains an additional section on generalized cores and stable sets, Section 7.3.

(5) Chapter 8 has a slightly modified structure in comparison to its corresponding chapter in the first edition, but its contents remained in fact unchanged.

(6) Chapter 10 is an enlarged version of Chapter 9 of the first edition due to the enhancing and enlarging of Chapters 11 and 12, where we use improved notation as well.

(7) Chapter 11, dealing with solution concepts for multi-choice games, has a new structure as to better incorporate the new text regarding Shapley-like values. Section 11.4 is new and contains the multi-choice version of the equal split-off set for crisp games (cf. Section 4.2).

(8) Chapter 12, originated from Chapter 11 of the first edition, contains several new notions and results on three classes of multi-choice games: balanced, convex and clan games. In particular, Section 12.1.2 introduces the notion of (level-increase) monotonic allocation schemes for whose existence the total balancedness of a multi-choice game is a necessary condition. Section 12.2 is almost entirely new. Specifically, Section 12.2.2 dealing with monotonic allocation schemes for convex multi-choice games is new, as well as Section 12.2.3 where the multi-choice version of the constrained egalitarian solution for convex crisp games (cf. Section 5.2.3) is introduced and studied, whereas Section 12.2.4 focuses on properties of all the foregoing solutions for convex games with multi-choice coalitions. Section 12.3 is entirely new. First, in Section 12.3.1 the notions of clan multi-choice game and total clan multi-choice games are introduced, and characterizations of total clan multi-choice games are provided. Second, in Section 12.3.2 we introduce and study bi-monotonic allocation schemes for a subclass of total clan multi-choice games, where suitably defined compensation-sharing rules play a key role.

We hope that this updated and enlarged edition will prove even more beneficial than the first edition for our readers. We express our gratitude to Katharina Wetzel-Vandai who encouraged us to prepare a second edition of our Springer 2005 book, and to Luis G. González Morales for transforming the manuscript into this final version.

Nijmegen,
January 2008

Rodica Branzei
Dinko Dimitrov
Stef Tijs

Preface to the First Edition

This book investigates the classical model of cooperative games with transferable utility (TU-games) and models in cooperative game theory in which the players have the possibility to cooperate partially, namely fuzzy games and multi-choice games. In a cooperative TU-game the agents are either fully involved or not involved at all in cooperation with some other agents, while in a fuzzy game players are allowed to cooperate with infinitely many different participation levels, varying from non-cooperation to full cooperation. A multi-choice game describes an "intermediate" case in the sense that each player may have a fixed finite number of activity levels.

Part I of the book is devoted to the most developed model in the theory of cooperative games, that of cooperative games in characteristic function form or cooperative games with transferable utility (TU-games), which we call here cooperative games with crisp coalitions or, simply, crisp games. It presents basic notions, solutions concepts and classes of cooperative crisp games in such a way that it allows the reader to use this part as a reference toolbox when studying the corresponding concepts from the theory of fuzzy games (Part II) and from the theory of multi-choice games (Part III).

The work on this book started while we were research fellows at ZiF (Bielefeld) for the project "Procedural Approaches to Conflict Resolution", 2002. We thank our hosts Matthias Raith and Olaf Gaus for giving us the possibility to freely structure our research plans and also the officials from the ZiF administration for their kind hospitality. The work of Dinko Dimitrov was generously supported by a Marie Curie Research Fellowship of the European Community programme "Improving the Human Research Potential and the Socio-Economic Knowledge Base" under contract number HPMF-CT-2002-02121 conducted at Tilburg University. Thanks are also due to Luis G. González Morales for transforming the manuscript into this final version.

Tilburg,
May 2005

Rodica Branzei
Dinko Dimitrov
Stef Tijs

Contents

Part III Multi-Choice Games

Part I

Cooperative Games with Crisp Coalitions

Cooperative game theory is concerned primarily with coalitions - groups of players - who coordinate their actions and pool their winnings. Consequently, one of the problems here is how to divide the extra earnings (or cost savings) among the members of the formed coalition. The basis of this theory was laid by John von Neumann and Oskar Morgenstern in [78] with coalitional games in characteristic function form, known also as transferable utility games (TU-games). Since then various solution concepts for cooperative TU-games have been proposed and several interesting subclasses of TU-games have been introduced. In what follows in this part we present a selection of basic notions, solution concepts and classes of cooperative TU-games that will be extensively used in the next two parts of the book. More detailed introductory books on the theory of (cooperative) games include [86] and [110], where also non-transferable utility games (NTU-games) are treated. The list can be enlarged by the very recent monographs [79], [87], [91], [123].

This part of the book is devoted to the most developed model in the theory of cooperative games, that of cooperative games in characteristic function form or cooperative games with transferable utility (TU-games), which we call here cooperative games with crisp coalitions or, simply, crisp games. It is organized as follows. Chapter 1 introduces basic notation, definitions and notions from cooperative game theory dealing with TU-games. In Chapter 2 we consider set-valued solution concepts like the core, the dominance core and stable sets, and different core catchers. The relations among these solution concepts are extensively studied. Chapter 3 is devoted to two well known one-point solutions concepts - the Shapley value and the τ-value, and to the average lexicographic value recently introduced in [111]. We present different formulations of these values, discuss some of their properties and axiomatic characterizations. In Chapter 4 we present an overview of egalitarianism-based solution concepts and introduce the equal split-off set for cooperative games, while in Chapter 5 we study three classes of cooperative games with crisp coalitions - totally balanced games, convex games and clan games. We discuss specific properties of the solution concepts introduced in Chapters 3 and 4 on these classes of games and present specific solution concepts like the concept of a population monotonic allocation scheme for totally balanced games, the constrained egalitarian solution for convex games, the concept of a bi-monotonic allocation scheme for clan games.

1 Preliminaries

Let N be a non-empty finite set of agents who consider different cooperation possibilities. Each subset $S \subset N$ is referred to as a *crisp coalition*. The set N is called the *grand coalition* and $\emptyset$ is called the *empty coalition*. We denote the collection of coalitions, i.e. the set of all subsets of N by 2^N. For each $S \in 2^N$ we denote by $|S|$ the number of elements of S, and by e^S the characteristic vector of S with $\left(e^S\right)^i = 1$ if $i \in S$, and $\left(e^S\right)^i = 0$ if $i \in N \setminus S$. In the following often $N = \{1, \ldots, n\}$.

Definition 1.1. *A* ***cooperative game*** *in characteristic function form is an ordered pair $\langle N, v \rangle$ consisting of the player set N and the characteristic function $v : 2^N \to \mathbb{R}$ with $v(\emptyset) = 0$.*

The real number $v(S)$ can be interpreted as the maximal worth or cost savings that the members of S can obtain when they cooperate. Often we identify the game $\langle N, v \rangle$ with its characteristic function v.

A cooperative game in characteristic function form is usually referred to as a transferable utility game (TU-game). A cooperative game might be a non-transferable utility game (NTU-game); the reader is referred to [87] and [110] for an introduction to NTU-games.

Example 1.2. (Glove game) Let $N = \{1, \ldots, n\}$ be divided into two disjoint subsets L and R. Members of L possess a left hand glove, members of R a right hand glove. A single glove is worth nothing, a right-left pair of gloves has value of one euro. This situation can be modeled as a game $\langle N, v \rangle$, where for each $S \in 2^N$ we have $v(S) := \min\{|L \cap S|, |R \cap S|\}$.

The set G^N of characteristic functions of coalitional games with player set N forms with the usual operations of addition and scalar multiplication of functions a $(2^{|N|}-1)$-dimensional linear space; a basis of this space is supplied by the *unanimity games* u_T, $T \in 2^N \setminus \{\emptyset\}$, that are defined by

$$u_T(S) = \begin{cases} 1 \text{ if } T \subset S, \\ 0 \text{ otherwise.} \end{cases} \tag{1.1}$$

One can easily check that for each $v \in G^N$ we have

$$v = \sum_{T \in 2^N \setminus \{\emptyset\}} c_T u_T \text{ with } c_T = \sum_{S: S \subset T} (-1)^{|T|-|S|} v(S). \tag{1.2}$$

The interpretation of the unanimity game u_T is that a gain (or cost savings) of 1 can be obtained if and only if all players in coalition S are involved in cooperation.

Definition 1.3. *A game $v \in G^N$ is called **simple**[1] if $v(S) \in \{0,1\}$ for all $S \in 2^N \setminus \{\emptyset\}$ and $v(\emptyset) = 0$, $v(N) = 1$.*

Note that the unanimity game u_T, $T \in 2^N \setminus \{\emptyset\}$, is a special simple game.

Definition 1.4. *A coalition S is **winning** in the simple game $v \in G^N$ if $v(S) = 1$.*

Definition 1.5. *A coalition S is **minimal winning** in the simple game $v \in G^N$ if $v(S) = 1$ and $v(T) = 0$ for all $T \subset S, T \neq S$.*

Definition 1.6. *A player $i \in N$ is a **dictator** in the simple game $v \in G^N$ if the coalition $\{i\}$ is minimal winning and there are no other minimal winning coalitions.*

Definition 1.7. *Let $v \in G^N$. For each $i \in N$ and for each $S \in 2^N$ with $i \in S$, the **marginal contribution** of player i to the coalition S is $M_i(S,v) := v(S) - v(S \setminus \{i\})$.*

Let $\pi(N)$ be the set of all permutations $\sigma : N \to N$ of N. The set $P^\sigma(i) := \{r \in N | \sigma^{-1}(r) < \sigma^{-1}(i)\}$ consists of all predecessors of i with respect to the permutation σ.

Definition 1.8. *Let $v \in G^N$ and $\sigma \in \pi(N)$. The **marginal contribution vector** $m^\sigma(v) \in \mathbb{R}^n$ with respect to σ and v has the i-th coordinate $m_i^\sigma(v) := v(P^\sigma(i) \cup \{i\}) - v(P^\sigma(i))$ for each $i \in N$.*

[1] In some game theory literature a game is simple if it is additionally monotonic (cf. Definition 1.11).

In what follows, we often write m^σ instead of $m^\sigma(v)$ when it is clear which game v we have in mind.

Definition 1.9. *For a game $v \in G^N$ and a coalition $T \in 2^N \setminus \{\emptyset\}$, the* ***subgame*** *with player set T is the game v_T defined by $v_T(S) := v(S)$ for all $S \in 2^T$.*

Hence, v_T is the restriction of v to the set 2^T.

Definition 1.10. *A game $v^* \in G^N$ is called the* ***dual game*** *of $v \in G^N$ if $v^*(S) = v(N) - v(N \setminus S)$ for all $S \subseteq N$.*

Definition 1.11. *A game $v \in G^N$ is said to be* ***monotonic*** *if $v(S) \leq v(T)$ for all $S, T \in 2^N$ with $S \subset T$.*

Definition 1.12. *A game $v \in G^N$ is called* ***non-negative*** *if for each $S \in 2^N$ we have $v(S) \geq 0$.*

Definition 1.13. *A game $v \in G^N$ is* ***additive*** *if $v(S \cup T) = v(S) + v(T)$ for all $S, T \in 2^N$ with $S \cap T = \emptyset$.*

An additive game $v \in G^N$ is determined by the vector

$$a = (v(\{1\}), ..., v(\{n\})) \in \mathbb{R}^n \tag{1.3}$$

since $v(S) = \sum_{i \in S} a_i$ for all $S \in 2^N$. Additive games form an n-dimensional linear subspace of G^N. A game $v \in G^N$ is called *inessential* if it is an additive game. For an inessential game there is no problem how to divide $v(N)$ because $v(N) = \sum_{i \in N} v(\{i\})$ (and also $v(S) = \sum_{i \in S} v(\{i\})$ for all $S \subset N)$[2].

Most of the cooperative games arising from real life situations are superadditive games.

Definition 1.14. *A game $v \in G^N$ is* ***superadditive*** *if $v(S \cup T) \geq v(S) + v(T)$ for all $S, T \in 2^N$ with $S \cap T = \emptyset$.*

Definition 1.15. *A game $v \in G^N$ is* ***subadditive*** *if $-v$ is superadditive.*

Of course, in a superadditive game we have $v\left(\cup_{i=1}^k S_i\right) \geq \sum_{i=1}^k v(S_i)$ if $S_1, \ldots, S_k$ are pairwise disjoint coalitions. Especially $v(N) \geq \sum_{i=1}^k v(S_i)$ for each partition $(S_1, \ldots, S_k)$ of N; in particular $v(N) \geq \sum_{i=1}^n v(i)$. Note that the game in Example 1.2 is superadditive. In a

[2] Given a game $v \in G^N$ and a coalition $\{i, \ldots, k\} \subset N$, we will often write $v(i, \ldots, k)$ instead of $v(\{i, \ldots, k\})$.

superadditive game it is advantageous for the players to cooperate. The set of (characteristic functions of) superadditive games form a *cone* in G^N, i.e. for all v and w that are superadditive we have that $\alpha v + \beta w$ is also a superadditive game, where $\alpha, \beta \in \mathbb{R}_+$.

Definition 1.16. *A game $v \in G^N$ for which $v(N) > \sum_{i=1}^{n} v(i)$ is said to be an **N-essential game**.*

In what follows in Part I, the notion of a *balanced game* will play an important role.

Definition 1.17. *A map $\lambda : 2^N \setminus \{\emptyset\} \to \mathbb{R}_+$ is called a **balanced map** if $\sum_{S \in 2^N \setminus \{\emptyset\}} \lambda(S)\, e^S = e^N$.*

Definition 1.18. *A collection B of coalitions is called **balanced** if there is a balanced map λ such that $B = \left\{S \in 2^N \mid \lambda(S) > 0\right\}$.*

Definition 1.19. *A game $v \in G^N$ is **balanced** if for each balanced map $\lambda : 2^N \setminus \{\emptyset\} \to \mathbb{R}_+$ we have*

$$\sum_{S \in 2^N \setminus \{\emptyset\}} \lambda(S)\, v(S) \leq v(N)\,. \tag{1.4}$$

Let us consider now two games $v, w \in G^N$ and answer the question "When can we say that v and w are 'essentially' the same?"

Definition 1.20. *Let $v, w \in G^N$. The game w is **strategically equivalent** to the game v if there exist $k > 0$ and an additive game a (cf. (1.3)) such that $w(S) = kv(S) + \sum_{i \in S} a_i$ for all $S \in 2^N \setminus \{\emptyset\}$.*

One may think that w arises out of v by the following changes:

- the unit of payoffs is changed, where the exchange rate is k;
- in the game w each player is given either a bonus (if $a_i > 0$) or a fee (if $a_i < 0$) before the distribution of $kv(N)$ among the players starts.

Notice that the strategic equivalence is an equivalence relation on the set G^N, i.e. we have:

- (Reflexivity) The game v is strategically equivalent to itself (take $k = 1$ and $a_i = 0$ for each $i \in N$);
- (Symmetry) If w is strategically equivalent to v, then v is strategically equivalent to w (if for all coalitions $S \subset N$, $w(S) = kv(S) + \sum_{i \in S} a_i$, then $v(S) = \frac{1}{k} w(S) - \sum_{i \in S} \frac{a_i}{k}$ and $\frac{1}{k} > 0$);

- (Transitivity) If w is strategically equivalent to v and u is strategically equivalent to w, then u is strategically equivalent to v ($w(S) = kv(S) + a(S)$ and $u(S) = lw(S) + b(S)$ imply $u(S) = lkv(S) + (la(S) + b(S))$, where $a(S) := \sum_{i \in S} a_i$).

For most solution concepts – as we will see later – it is sufficient to look only at one of the games in an (strategic) equivalence class. One considers often games in an equivalence class that are in (α, β)-form for $\alpha, \beta \in \mathbb{R}$.

Definition 1.21. *Let $\alpha, \beta \in \mathbb{R}$. A game $v \in G^N$ is called a **game in (α, β)-form** if $v(i) = \alpha$ for all $i \in N$ and $v(N) = \beta$.*

Theorem 1.22. *Each N-essential game $v \in G^N$ is strategically equivalent to a game $w \in G^N$ in $(0,1)$-form. This game is unique.*

Proof. For some $k > 0$ and $a_1, ..., a_n \in \mathbb{R}$ we try to find a game w with $w(S) = kv(S) + a(S)$ for all $S \in 2^N \setminus \{\emptyset\}$, $w(\{i\}) = 0$ for all $i \in N$, and $w(N) = 1$. Then necessarily

$$w(i) = 0 = kv(i) + a_i, \tag{1.5}$$

$$w(N) = 1 = kv(N) + \sum_{i \in N} a_i. \tag{1.6}$$

Then $w(N) - \sum_{i \in N} w(i) = 1 = k\left(v(N) - \sum_{i \in N} v(i)\right)$ by (1.5) and (1.6). Hence, $k = \frac{1}{v(N) - \sum_{i \in N} v(i)}$. From (1.5) we derive $a_i = -\frac{v(i)}{v(N) - \sum_{i \in N} v(i)}$. If we take for all $S \in 2^N \setminus \{\emptyset\}$, $w(S) = \frac{v(S) - \sum_{i \in S} v(i)}{v(N) - \sum_{i \in N} v(i)}$, then we obtain the unique game w in $(0,1)$-form, which is strategically equivalent to v.

Definition 1.23. *A game $v \in G^N$ is called **zero-normalized** if for all $i \in N$ we have $v(i) = 0$.*

One can easily check that each game $v \in G^N$ is strategically equivalent to a zero-normalized game $w \in G^N$, where $w(S) = v(S) - \sum_{i \in S} v(i)$.

Definition 1.24. *A game $v \in G^N$ is said to be **zero-monotonic** if its zero-normalization is monotonic.*

It holds that a game which is strategically equivalent to a zero-monotonic game is also zero-monotonic.

We turn now to one of the basic questions in the theory of cooperative TU-games: "If the grand coalition forms, how to divide the profit or cost savings $v(N)$?"

This question is approached with the aid of solution concepts in cooperative game theory like cores, stable sets, bargaining sets, the Shapley value, the τ-value, the nucleolus. A solution concept gives an answer to the question of how the reward (cost savings) obtained when all players in N cooperate should be distributed among the individual players while taking account of the potential reward (cost savings) of all different coalitions of players. Hence, a solution concept assigns to a coalitional game at least one payoff vector $x = (x_i)_{i \in N} \in \mathbb{R}^n$, where x_i is the payoff allocated to player $i \in N$. A selection of (set-valued and one-point) solution concepts which will be used along this book, their axiomatic characterizations and interrelations will be given in Chapters 2-4.

Definition 1.25. *A* ***set-valued solution*** *(or a multisolution) is a multifunction $F : G^N \to\to \mathbb{R}^n$.*

Definition 1.26. *An* ***one-point solution*** *(or a single-valued rule) is a map $f : G^N \to \mathbb{R}^n$.*

We mention now some desirable properties for one-point solution concepts. Extensions of these properties to set-valued solution concepts are straightforward.

Definition 1.27. *Let $f : G^N \to \mathbb{R}^n$. Then f satisfies*
(i) ***individual rationality*** *if $f_i(v) \geq v(i)$ for all $v \in G^N$ and $i \in N$.*
(ii) ***efficiency*** *if $\sum_{i=1}^n f_i(v) = v(N)$ for all $v \in G^N$.*
(iii) ***relative invariance with respect to strategic equivalence*** *if for all $v, w \in G^N$, all additive games $a \in G^N$, and all $k > 0$ we have that $w = kv + a$ implies $f(kv + a) = kf(v) + a$.*
(iv) the ***dummy player property*** *if $f_i(v) = v(i)$ for all $v \in G^N$ and for all dummy players i in v, i.e. players $i \in N$ such that $v(S \cup \{i\}) = v(S) + v(i)$ for all $S \in 2^{N \setminus \{i\}}$.*
(v) the ***anonymity property*** *if $f(v^\sigma) = \sigma^*(f(v))$ for all $\sigma \in \pi(N)$. Here v^σ is the game with $v^\sigma(\sigma(U)) := v(U)$ for all $U \in 2^N$ or $v^\sigma(S) = v(\sigma^{-1}(S))$ for all $S \in 2^N$ and $\sigma^* : \mathbb{R}^n \to \mathbb{R}^n$ is defined by $(\sigma^*(x))_{\sigma(k)} := x_k$ for all $x \in \mathbb{R}^n$ and $k \in N$.*
(vi) ***additivity*** *if $f(v + w) = f(v) + f(w)$ for all $v, w \in G^N$.*

We end this chapter by recalling some definitions and results from linear algebra which are used later.

Definition 1.28. *Let V and W be vector spaces over $\mathbb{R}$. Let $L : V \to W$ be a map. Then L is called a* **linear transformation** *(linear map, linear operator) from V into W if for all $x, y \in V$ and all $\alpha, \beta \in \mathbb{R}$ we have $L(\alpha x + \beta y) = \alpha L(x) + \beta L(y)$.*

Definition 1.29. *A set W is a (linear)* **subspace** *of the vector space V if $W \subset V$, $0 \in W$, and W is closed with respect to addition and scalar multiplication, i.e. for all $x, y \in W$ we have $x + y \in W$, and for each $x \in W$ and $\alpha \in \mathbb{R}$, also $\alpha x \in W$ holds.*

Definition 1.30. *A subset C of a vector space V over $\mathbb{R}$ is called* **convex** *if for all $x, y \in C$ and all $\alpha \in (0, 1)$ we have $\alpha x + (1 - \alpha) y \in C$.*

A geometric interpretation of a convex set is that with each pair x, y of points in it, the line segment with x, y as endpoints is also in the set.

Definition 1.31. *Let C be a convex set. A point $x \in C$ is called an* **extreme point** *of C if there do not exist $x_1, x_2 \in C$ with $x_1 \neq x$, $x_2 \neq x$ and $\alpha \in (0, 1)$ such that $x = \alpha x_1 + (1 - \alpha) x_2$. The set of extreme points of a convex set C will be denoted by* $\mathrm{ext}(C)$.

Definition 1.32. *A set H of points in $\mathbb{R}^n$ is called a* **hyperplane** *if it is the set of solutions of a linear equation $a_1 x_1 + ... + a_n x_n = b$, with $(a_1, ..., a_n) \in \mathbb{R}^n \setminus \{0\}$ and $b \in \mathbb{R}$. A hyperplane separates a (linear) space in two (linear) halfspaces. Let A be an $n \times p$ matrix and $b \in \mathbb{R}^p$; a set $P = \left\{x \in \mathbb{R}^n \mid x^T A \geq b^T\right\}$ is called a* **polyhedral set**.

The following theorem gives a characterization of extreme points of a polyhedral set.

Theorem 1.33. *Let A be an $n \times p$ matrix, $b \in \mathbb{R}^p$ and let P be the polyhedral set of solutions of the set of inequalities $x^T A \geq b^T$. For $x \in \mathbb{R}^n$ let* $\mathrm{tight}(x)$ *be the set of columns $\left\{Ae^j \mid x^T A e^j = b_j\right\}$ of A where the corresponding inequalities are equalities for x, and where for each $j \in N$, e^j is the j-th standard basis vector in $\mathbb{R}^n$. Then x is an extreme point of P iff* $\mathrm{tight}(x)$ *is a complete system of vectors in $\mathbb{R}^n$.*

The next theorem is known as the duality theorem from linear programming theory.

Theorem 1.34. *Let A be an $n \times p$ matrix, $b \in \mathbb{R}^p$ and $c \in \mathbb{R}^n$. Then* $\min\left\{x^T c \mid x^T A \geq b^T\right\} = \max\left\{b^T y \mid Ay = c, y \geq 0\right\}$ *if* $\left\{x \in \mathbb{R}^n \mid x^T A \geq b^T\right\} \neq \emptyset$ *and* $\{y \in \mathbb{R}^p \mid Ay = c, y \geq 0\} \neq \emptyset$.

Definition 1.35. *Let V be a vector space and $A \subset V$. The* ***convex hull*** $\operatorname{co}(A)$ *of A is the set*

$$\left\{x \in V \mid \exists p \in \mathbb{N}, \alpha \in \Delta^p, v_1, ..., v_p \in A \text{ s.t. } \sum_{i=1}^{p} \alpha_i v_i = x\right\},$$

where $\Delta^p = \left\{q \in \mathbb{R}_+^p \mid \sum_{i=1}^p q_i = 1\right\}$ *is the $(p-1)$ dimensional unit simplex.*

2

Cores and Related Solution Concepts

In this chapter we consider payoff vectors $x = (x_i)_{i\in N} \in \mathbb{R}^n$, with x_i being the payoff to be given to player $i \in N$, under the condition that cooperation in the grand coalition is reached. Clearly, the actual formation of the grand coalition is based on the agreement of all players upon a proposed payoff in the game. Such an agreement is, or should be, based on all other cooperation possibilities for the players and their corresponding payoffs.

2.1 Imputations, Cores and Stable Sets

We note first that only payoff vectors $x \in \mathbb{R}^n$ satisfying $\sum_{i\in N} x_i \leq v(N)$ are reachable in the game $v \in G^N$ and the set of such payoff vectors is nonempty and convex. More precisely, it is a halfspace of $\mathbb{R}^n$. We denote this set by $I^{**}(v)$, i.e.

$$I^{**}(v) := \left\{ x \in \mathbb{R}^n \mid \sum_{i\in N} x_i \leq v(N) \right\}.$$

However, to have any chance of being agreed upon, a payoff vector should satisfy *efficiency*, i.e.

$$\sum_{i\in N} x_i = v(N).$$

To motivate the efficiency condition we argue that $\sum_{i\in N} x_i \geq v(N)$ should also hold.

Suppose that $\sum_{i\in N} x_i < v(N)$. In this case we would have

$$a = v(N) - \sum_{i \in N} x_i > 0.$$

Then the players can still form the grand coalition and receive the better payoff $y = (y_1, \ldots, y_n)$ with $y_i = x_i + \frac{a}{n}$ for all $i \in N$.

We denote by $I^*(v)$ the set of efficient payoff vectors in the coalitional game $v \in G^N$, i.e.

$$I^*(v) := \left\{ x \in \mathbb{R}^n \mid \sum_{i \in N} x_i = v(N) \right\}$$

and, clearly, $I^*(v) \neq \emptyset$. This convex set is referred to as the *preimputation set* of the game $v \in G^N$. It is a hyperplane in $\mathbb{R}^n$. Clearly, $I^*(v) \subset I^{**}(v)$.

Now, note that if the proposed allocation $x \in I^*(v)$ is such that there is at least one player $i \in N$ whose payoff x_i satisfies $x_i < v(i)$, the grand coalition would never form. The reason is that such a player would prefer not to cooperate since acting on his own he can obtain more.

Hence, the *individual rationality* condition

$$x_i \geq v(i) \text{ for all } i \in N$$

should hold in order that a payoff vector has a real chance to be realized in the game.

Definition 2.1. *A payoff vector $x \in \mathbb{R}^n$ is an* ***imputation*** *for the game $v \in G^N$ if it is efficient and individually rational, i.e.*
(i) $\sum_{i \in N} x_i = v(N)$;
(ii) $x_i \geq v(i)$ for all $i \in N$.

We denote by $I(v)$ the set of imputations of $v \in G^N$. Clearly, $I(v)$ is empty if and only if $v(N) < \sum_{i \in N} v(i)$. Further, for an additive game (cf. Definition 1.13),

$$I(v) = \{(v(1), \ldots, v(n))\}.$$

The next theorem shows that N-essential games (cf. Definition 1.16) always have infinitely many imputations. Moreover, $I(v)$ is a simplex with extreme points $f^1(v), \ldots, f^n(v)$, where for each $i \in N$, $f^i(v) = \left(f_1^i(v), \ldots, f_j^i(v), \ldots, f_n^i(v)\right)$ with

$$f_j^i(v) = \begin{cases} v(i) & \text{if } i \neq j, \\ v(N) - \sum_{k \in N \setminus \{i\}} v(k) & \text{if } i = j. \end{cases} \tag{2.1}$$

Theorem 2.2. *Let $v \in G^N$. If v is N-essential, then*
(i) $I(v)$ is an infinite set.
(ii) $I(v)$ is the convex hull of the points $f^1(v), \dots, f^n(v)$.

Proof. (i) Since $v \in G^N$ is an N-essential game we have $a = v(N) - \sum_{i \in N} x_i > 0$. For any n-tuple $b = (b_1, \dots, b_n)$ of nonnegative numbers such that $\sum_{i \in N} b_i = a$, the payoff vector $x' = (x'_1, \dots, x'_n)$ with $x'_i = v(i) + b_i$ for all $i \in N$ is an imputation.

(ii) This follows from Theorem 1.33 by noting that

$$I(v) = \left\{ x \in \mathbb{R}^n \mid x^T A \geq b^T \right\},$$

where A is the $n \times (n+2)$-matrix with columns $e^1, \dots, e^n, 1^n, -1^n$ and

$$b = (v(1), \dots, v(n), v(N), -v(N)),$$

where for each $i \in N$, e^i is the i-th standard basis in $\mathbb{R}^n$ and 1^n is the vector in $\mathbb{R}^n$ with all coordinates equal to 1.

Since the imputation set of an N-essential game is too large according to the above theorem, there is a need for some criteria to single out those imputations that are most likely to occur. In this way one obtains subsets of $I(v)$ as *solution concepts.*

The first (set-valued) solution concept we would like to study is the *core* of a game (cf. [52]).

Definition 2.3. *The* **core** *$C(v)$ of a game $v \in G^N$ is the set*

$$\left\{ x \in I(v) \mid \sum_{i \in S} x_i \geq v(S) \text{ for all } S \in 2^N \setminus \{\emptyset\} \right\}.$$

If $x \in C(v)$, then no coalition S has an incentive to split off if x is the proposed reward allocation in N, because the total amount $\sum_{i \in S} x_i$ allocated to S is not smaller than the amount $v(S)$ which the players can obtain by forming the subcoalition. If $C(v) \neq \emptyset$, then elements of $C(v)$ can easily be obtained because the core is defined with the aid of a finite system of linear inequalities. The core is a polytope.

In [16] and [103] one can find a characterization of games with a nonempty core that we present in the next theorem.

Theorem 2.4. *Let $v \in G^N$. Then the following two assertions are equivalent:*
(i) $C(v) \neq \emptyset$,
(ii) The game v is balanced (cf. Definition 1.19).

Proof. First we note that $C(v) \neq \emptyset$ iff

$$v(N) = \min\left\{\sum_{i\in N} x_i \mid \sum_{i\in S} x_i \geq v(S) \text{ for all } S \in 2^N \setminus \{\emptyset\}\right\}. \quad (2.2)$$

By Theorem 1.34, equality (2.2) holds iff

$$v(N) = \max\left\{\sum_{S\in 2^N\setminus\{\emptyset\}} \lambda(S)v(S) \mid \sum_{S\in 2^N\setminus\{\emptyset\}} \lambda(S)e^S = e^N, \lambda \geq 0\right\} \quad (2.3)$$

(take for A the matrix with the characteristic vectors e^S as columns). Now, (2.3) holds iff (1.4) holds. Hence, (i) and (ii) are equivalent.

Now, we reformulate this Bondareva-Shapley result in geometric terms (cf. [26]). Let for a subsimplex $\Delta(S,v) = \operatorname{conv}\{f^i(v) \mid i \in S\}$ of $I(v) = \Delta(N,v)$, the barycenter $\frac{1}{|S|}\sum_{i\in S} f^i(v)$ be denoted by $b(S,v)$. Then we obtain the following characterization of games with a non-empty core.

Theorem 2.5. *A game $v \in G^N$ has a non-empty core iff*

$$\sum_{S\in 2^N\setminus\{\emptyset\}} \mu_S b(S,v) = b(N,v)$$

with $\mu_S \geq 0$ and $\sum_{S\in 2^N\setminus\{\emptyset\}} \mu_S = 1$ implies $\sum_{S\in 2^N\setminus\{\emptyset\}} \mu_S \frac{v(S)}{|S|} \leq \frac{v(N)}{|N|}$.

Proof. For $\lambda = (\lambda_S)_{S\in 2^N\setminus\{\emptyset\}}$, let $\mu = (\mu_S)_{S\in 2^N\setminus\{\emptyset\}}$ be defined by $\mu_S = n^{-1}|S|\lambda_S$. Then

$$\lambda_S \geq 0, \quad \sum_{S\in 2^N\setminus\{\emptyset\}} \lambda_S e^S = e^N \text{ iff}$$

$$\sum_{S\in 2^N\setminus\{\emptyset\}} \mu_S \frac{e^S}{|S|} = \frac{e^N}{|N|}, \ \mu_S \geq 0, \ \sum_{S\in 2^N\setminus\{\emptyset\}} \mu_S = 1.$$

This implies

(i) $\lambda_S \geq 0$, $\sum_{S\in 2^N\setminus\{\emptyset\}} \lambda_S e^S = e^N$ iff $\sum_{S\in 2^N\setminus\{\emptyset\}} \mu_S b(S,v) = b(N,v)$, since $b(S,v) = (v(\{1\}), v(\{2\}), ..., v(\{n\})) + \alpha|S|^{-1}e^S$for each $S \in 2^N \setminus \{\emptyset\}$, where $\alpha = v(N) - \sum_{i\in N} v(\{i\})$.

(ii) $\sum_{S\in 2^N\setminus\{\emptyset\}} \lambda_S v(S) \leq v(N)$ iff $\sum_{S\in 2^N\setminus\{\emptyset\}} \mu_S \frac{v(S)}{|S|} \leq \frac{v(N)}{|N|}$.

The theorem tells us that a game $v \in G^N$ has a non-empty core if and only if for each way of writing the barycenter of the imputation set as a convex combination of barycenters of subsimplices, the *per capita value* $v(N)/|N|$ of N is at least as large as the corresponding convex combination of the per capita values of the subcoalitions $(v(S)/|S|)$.

Remark 2.6. The core is relative invariant with respect to strategic equivalence (cf. Definition 1.27(iii)): if $w \in G^N$ is strategically equivalent to $v \in G^N$, say $w = kv + a$, then

$$C(w) = kC(v) + a \,(:= \{x \in \mathbb{R}^n \mid x = ky + a \text{ for some } y \in C(v)\}) .$$

For axiomatic characterizations of the core, we refer the reader to [67], [85], and [87].

Other subsets of imputations which are solution concepts for coalitional games are the *dominance core* (D-core) and *stable sets* (cf. [78]). They are defined based on the following dominance relation over vectors in $\mathbb{R}^n$.

Definition 2.7. *Let $v \in G^N$, $x, y \in I(v)$, and $S \in 2^N \setminus \{\emptyset\}$. We say that x **dominates** y **via coalition** S, and denote it by $x \operatorname{dom}_S y$ if*
(i) $x_i > y_i$ for all $i \in S$,
(ii) $\sum_{i \in S} x_i \leq v(S)$.

Note that if (i) holds, then the payoff x is better than the payoff y for all members of S; condition (ii) guarantees that the payoff x is reachable for S.

Definition 2.8. *Let $v \in G^N$ and $x, y \in I(v)$. We say that x **dominates** y, and denote it by $x \operatorname{dom} y$ if there exists $S \in 2^N \setminus \{\emptyset\}$ such that $x \operatorname{dom}_S y$.*

Proposition 2.9. *Let $v \in G^N$ and $S \in 2^N \setminus \{\emptyset\}$. Then the relations* dom_S *and* dom *are irreflexive. Moreover,* dom_S *is transitive and antisymmetric.*

Proof. That dom_S and dom are irreflexive follows from the fact that for $x \in I(v)$ there is no $S \in 2^N \setminus \{\emptyset\}$ such that $x_i > x_i$ for all $i \in S$.

To prove that dom_S is transitive take $x, y, z \in I(v)$ such that $x \operatorname{dom}_S y$ and $y \operatorname{dom}_S z$. Then $x_i > z_i$ for all $i \in S$. So $x \operatorname{dom}_S z$.

To prove that dom_S is antisymmetric, suppose $x \operatorname{dom}_S y$. Then $x_i > y_i$ for all $i \in S$, i.e. there is no $i \in S$ such that $y_i > x_i$. Hence, $y \operatorname{dom}_S x$ does not hold.

For $S \in 2^N \setminus \{\emptyset\}$ we denote by $D(S)$ the set of imputations which are dominated via S; note that players in S can successfully protest against any imputation in $D(S)$.

Definition 2.10. *The* ***dominance core*** *(****D-core****) $DC(v)$ of a game $v \in G^N$ consists of all undominated elements in $I(v)$, i.e. the set $I(v) \setminus \bigcup_{S \in 2^N \setminus \{\emptyset\}} D(S)$.*

It turns out that $DC(v)$ is also a convex set; moreover, it is a polytope and relative invariant with respect to strategic equivalence. We refer to [67] for an axiomatic framework to compare the D-core and the core of a cooperative game.

For $v \in G^N$ and $A \subset I(v)$ we denote by $\mathrm{dom}\,(A)$ the set consisting of all imputations that are dominated by some element in A. Note that $DC(v) = I(v) \setminus \mathrm{dom}(I(v))$.

Definition 2.11. *For $v \in G^N$ a subset K of $I(v)$ is called a* ***stable set*** *if the following conditions hold:*
(i) (Internal stability) $K \cap \mathrm{dom}(K) = \emptyset$,
(ii) (External stability) $I(v) \setminus K \subset \mathrm{dom}(K)$.

This notion was introduced by von Neumann and Morgenstern (cf. [78]) with the interpretation that a stable set corresponds to a "standard of behavior", which, if generally accepted, is self-enforcing.

The two conditions in Definition 2.11 can be interpreted as follows:

- By external stability, an imputation outside a stable set K seems unlikely to become established: there is always a coalition that prefers one of the achievable imputations inside K, implying that there would exist a tendency to shift to an imputation in K;
- By internal stability, all imputations in K are "equal" with respect to the dominance relation via coalitions, i.e. there is no imputation in K that is dominated by another imputation in K.

Note that for a game $v \in G^N$ the set K is a stable set if and only if K and $\mathrm{dom}(K)$ form a partition of $I(v)$. In principle, a game may have many stable sets or no stable set.

Theorem 2.12. *Let $v \in G^N$ and K be a stable set of v. Then*
(i) $C(v) \subset DC(v) \subset K$;
(ii) If v is superadditive, then $DC(v) = C(v)$;
(iii) If $DC(v)$ is a stable set, then there is no other stable set.

Proof. (i) In order to show that $C(v) \subset DC(v)$, let us suppose that there is $x \in C(v)$ such that $x \notin DC(v)$. Then there is an $y \in I(v)$ and a coalition $S \in 2^N \setminus \{\emptyset\}$ such that $y \,\mathrm{dom}_S\, x$. Then $v(S) \geq \sum_{i \in S} y_i > \sum_{i \in S} x_i$ which implies that $x \notin C(v)$.

To prove next that $DC(v) \subset K$ it is sufficient to show that $I(v) \setminus K \subset I(v) \setminus DC(v)$. Take $x \in I(v) \setminus K$. By the external stability of K there is a $y \in K$ with $y \,\mathrm{dom}\, x$. The elements in $DC(v)$ are not dominated. So $x \notin DC(v)$, i.e. $x \in I(v) \setminus DC(v)$.

(ii) We divide the proof of this assertion into two parts.

(ii.1) We show that for an $x \in I(v)$ with $\sum_{i \in S} x_i < v(S)$ for some $S \in 2^N \setminus \{\emptyset\}$, there is $y \in I(v)$ such that $y \,\mathrm{dom}_S\, x$. Define y as follows. If $i \in S$, then $y_i := x_i + \frac{v(S) - \sum_{i \in S} x_i}{|S|}$. If $i \notin S$, then $y_i := v(i) + \frac{v(N) - v(S) - \sum_{i \in N \setminus S} v(i)}{|N \setminus S|}$. Then $y \in I(v)$, where for the proof of $y_i \geq v(i)$ for $i \in N \setminus S$ we use the superadditivity of the game. Furthermore, $y \,\mathrm{dom}_S\, x$.

(ii.2) In order to show $DC(v) = C(v)$ we have, in view of (i), only to prove that $DC(v) \subset C(v)$. Suppose $x \in DC(v)$. Then there is no $y \in I(v)$ such that $y \,\mathrm{dom}\, x$. In view of (ii.1) we then have $\sum_{i \in S} x_i \geq v(S)$ for all $S \in 2^N \setminus \{\emptyset\}$. Hence, $x \in C(v)$.

(iii) Suppose $DC(v)$ is a stable set. Let K also be stable. By (i) we have $DC(v) \subset K$. To prove $K = DC(v)$, we have to show that $K \setminus DC(v) = \emptyset$. Suppose, to the contrary, that there is $x \in K \setminus DC(v)$. By the external stability of $DC(v)$ there is $y \in DC(v)$ $(\subset K)$ such that $y \,\mathrm{dom}\, x$. This is a contradiction to the internal stability of K. Hence $K \setminus DC(v) = \emptyset$ holds.

In addition to the relations among the core, the dominance core and the stable sets as established in Theorem 2.12, we state next without proof some additional results that will be used in the next parts of the book.

Theorem 2.13. *Let $v \in G^N$. Then*
(i) If $DC(v) \neq \emptyset$ and $v(N) \geq v(S) + \sum_{i \in N \setminus S} v(i)$ for each $S \subset N$, then $C(v) = DC(v)$.
(ii) If $C(v) \neq DC(v)$, then $C(v) = \emptyset$.

For details with respect to these relations the reader is referred to [42], [90], [105], and [110].

Another core-like solution concept which is related to the norm of equity is the equal division core introduced in [98].

Definition 2.14. *The* ***equal division core*** $EDC(v)$ *of a game* $v \in G^N$ *is the set*

$$\left\{x \in I(v) \mid \nexists S \in 2^N \setminus \{\emptyset\} \text{ s.t. } \frac{v(S)}{|S|} > x_i \text{ for all } i \in S\right\}.$$

In other words, the equal division core of a game consists of efficient payoff vectors for the grand coalition which cannot be improved upon by the equal division allocation of any subcoalition. It is clear that the core of a cooperative game is included in the equal division core of that game. The reader can find axiomatic characterizations of this solution concept on two classes of cooperative games in [13].

2.2 The Core Cover, the Reasonable Set and the Weber Set

In this section we introduce three sets related to the core, namely the *core cover* (cf. [116]), the *reasonable set* (cf. [51], [69], and [72]), and the *Weber set* (cf. [124]). All these sets can be seen as "core catchers" in the sense that they all contain the core of the corresponding game as a subset.

In the definition of the core cover the *upper vector* $M(N,v)$ and the *lower vector* $m(v)$ of a game $v \in G^N$ play a role.

For each $i \in N$, the i-th coordinate $M_i(N,v)$ of the upper vector $M(N,v)$ is the marginal contribution of player i to the grand coalition (cf. Definition 1.7); it is also called the *utopia payoff* for player i in the grand coalition in the sense that if player i wants more, then it is advantageous for the other players in N to throw player i out.

Definition 2.15. *Let* $S \in 2^N \setminus \{\emptyset\}$ *and* $i \in S$. *The* ***remainder*** $R(S,i)$ *of player* i *in the coalition* S *is the amount which remains for player* i *if coalition* S *forms and all other players in* S *obtain their utopia payoffs, i.e.*

$$R(S,i) := v(S) - \sum_{j \in S \setminus \{i\}} M_j(N,v).$$

For each $i \in N$, the i-th coordinate $m_i(v)$ of the *lower vector* $m(v)$ is then defined by

$$m_i(v) := \max_{S: i \in S} R(S,i).$$

We refer to $m_i(v)$ also as the *minimum right payoff* for player i, since this player has a reason to ask at least $m_i(v)$ in the grand coalition

N, by arguing that he can obtain that amount also by drumming up a coalition S with $m_i(v) = R(S,i)$ and making all other players in S happy with their utopia payoffs.

Definition 2.16. *The* ***core cover*** $CC(v)$ *of* $v \in G^N$ *consists of all imputations which are between* $m(v)$ *and* $M(N,v)$ *(in the usual partial order of* $\mathbb{R}^n$*), i.e.*

$$CC(v) := \{x \in I(v) \mid m(v) \leq x \leq M(N,v)\}.$$

That $CC(v)$ is a core catcher follows from the following theorem, which tells us that the lower (upper) vector is a lower (upper) bound for the core.

Theorem 2.17. *Let* $v \in G^N$ *and* $x \in C(v)$. *Then* $m(v) \leq x \leq M(N,v)$ *i.e.* $m_i(v) \leq x_i \leq M_i(N,v)$ *for all* $i \in N$.

Proof. (i) $x_i = x(N) - x(N \setminus \{i\}) = v(N) - x(N \setminus \{i\}) \leq v(N) - v(N \setminus \{i\}) = M_i(N,v)$ for each $i \in N$.

(ii) In view of (i), for each $S \subset N$ and each $i \in S$ we have

$$x_i = x(S) - x(S \setminus \{i\}) \geq v(S) - \sum_{j \in S \setminus \{i\}} M_j(N,v) = R(S,i).$$

So, $x_i \geq \max_{S:i \in S} R(S,i) = m_i(v)$ for each $i \in S$.

Another core catcher for a game $v \in G^N$ is introduced (cf. [72]) as follows.

Definition 2.18. *The* ***reasonable set*** $R(v)$ *of a game* $v \in G^N$ *is the set*

$$\left\{x \in \mathbb{R}^n \mid v(i) \leq x_i \leq \max_{S:i \in S} \left(v(S) - v(S \setminus \{i\})\right)\right\}.$$

Obviously, $C(v) \subset CC(v) \subset R(v)$. For an axiomatic characterization of the reasonable set we refer to [51].

The last core catcher for a game $v \in G^N$ we introduce (cf. [124]) is the Weber set. In its definition the marginal contribution vectors (cf. Definition 1.8) play a role.

Definition 2.19. *The* ***Weber set*** $W(v)$ *of a game* $v \in G^N$ *is the convex hull of the* $n!$ *marginal vectors* $m^\sigma(v)$, *corresponding to the* $n!$ *permutations* $\sigma \in \pi(N)$.

Here $m^{\sigma}(v)$ is the vector with

$$\begin{aligned} m^{\sigma}_{\sigma(1)}(v) &:= v(\sigma(1)), \\ m^{\sigma}_{\sigma(2)}(v) &:= v(\sigma(1), \sigma(2)) - v(\sigma(1)), \\ &\dots \\ m^{\sigma}_{\sigma(k)}(v) &:= v(\sigma(1), \dots, \sigma(k)) - v(\sigma(1), \dots, \sigma(k-1)) \end{aligned}$$

for each $k \in N$. The payoff vector m^{σ} can be created as follows. Let the players enter a room one by one in the order $\sigma(1), \dots, \sigma(n)$ and give each player the marginal contribution he creates by entering.

The Weber set is a core catcher as shown in

Theorem 2.20. *Let $v \in G^N$. Then $C(v) \subset W(v)$.*

Proof. If $|N| = 1$, then $I(v) = C(v) = W(v) = \{(v(1))\}$.

For $|N| = 2$ we consider two cases: $I(v) = \emptyset$ and $I(v) \neq \emptyset$. If $I(v) = \emptyset$, then $C(v) \subset I(v) = \emptyset \subset W(v)$. If $I(v) \neq \emptyset$, then we let

$$x' = (v(1), v(1,2) - v(1))$$

and

$$x'' = (v(2), v(1,2) - v(2)),$$

and note that

$$\begin{aligned} C(v) = I(v) &= co\{x', x''\} \\ &= co\{m^{\sigma}(v) \mid \sigma : \{1,2\} \to \{1,2\}\} = W(v). \end{aligned}$$

We proceed by induction on the number of players. So, suppose $|N| = n > 2$ and suppose that the core is a subset of the Weber set for all games with number of players smaller than n.

Since $C(v)$ and $W(v)$ are convex sets we need only to show that $x \in ext(C(v))$ implies $x \in W(v)$. Take $x \in ext(C(v))$. Then it follows from Theorem 1.33 that there exists $T \in 2^N \setminus \{\emptyset, N\}$ with $x(T) = v(T)$. Consider the $|T|$-person game u and the $(n - |T|)$-person game w defined by

$$\begin{aligned} u(S) &= v(S) \text{ for each } S \in 2^T, \\ w(S) &= v(T \cup S) - v(T) \text{ for each } S \in 2^{N\setminus T}. \end{aligned}$$

Then, obviously, $x^T \in C(u)$, and also $x^{N\setminus T} \in C(w)$ because $x^{N\setminus T}(S) = x(S) = x(T \cup S) - x(T) \geq v(T \cup S) - x(T) = v(T \cup S) - v(T) = w(S)$ for all $S \in 2^{N\setminus T}$ and

$$\sum_{i \in N \setminus T} x_i^{N \setminus T} = x(N) - x(T) = v(N) - v(T) = w(N \setminus T).$$

Since $|T| < n$, $|N \setminus T| < n$, the induction hypothesis implies that $x^T \in W(u)$ and $x^{N \setminus T} \in W(w)$.

Then $x = x^T \times x^{N \setminus T} \in W(u) \times W(w) \subset W(v)$. This last inclusion can be seen as follows. The extreme points of $W(u) \times W(w)$ are of the form (m^ρ, m^τ), where $\rho : \{1, 2, \ldots, |T|\} \to T$ and $\tau : \{1, 2, \ldots, |N \setminus T|\} \to N \setminus T$ are bijections and $m^\rho \in \mathbb{R}^{|T|}, m^\tau \in \mathbb{R}^{|N \setminus T|}$ are given by

$m^\rho_{\rho(1)} := u(\rho(1)) = v(\rho(1))$,

$m^\rho_{\rho(2)} := u(\rho(1), \rho(2)) - u(\rho(1)) = v(\rho(1), \rho(2)) - v(\rho(1))$,

...

$m^\rho_{\rho(|T|)} := u(T) - u(T \setminus \{\rho(|T|)\}) = v(T) - v(T \setminus \{\rho(|T|)\})$,

$m^\tau_{\tau(1)} := w(\tau(1)) = v(T \cup \{\tau(1)\}) - v(T)$,

$m^\tau_{\tau(2)} := w(\tau(1), \tau(2)) - w(\tau(1)) = v(T \cup \{\tau(1), \tau(2)\}) - v(T \cup \{\tau(1)\})$,

...

$m^\tau_{\tau(|N \setminus T|)} := w(N \setminus T) - w((N \setminus T) \setminus \{\tau(|N \setminus T|)\}) = v(N) - v(N \setminus \{\tau(|N \setminus T|)\})$.

Hence, $(m^\rho, m^\tau) \in \mathbb{R}^n$ corresponds to the marginal vector m^σ in $W(v)$ where $\sigma : N \to N$ is defined by

$$\sigma(i) := \begin{cases} \rho(i) & \text{if } 1 \le i \le |T|, \\ \tau(i - |T|) & \text{if } |T| + 1 \le i \le n. \end{cases}$$

So, we have proved that $ext(W(u) \times W(w)) \subset W(v)$. Since $W(v)$ is convex, $W(u) \times W(w) \subset W(v)$. We have proved that $x \in ext(C(v))$ implies $x \in W(v)$.

For another proof of Theorem 2.20 the reader is referred to [38].

3

The Shapley Value, the τ-value, and the Average Lexicographic Value

The Shapley value, the τ-value and the average lexicographic value recently introduced in [111] are three interesting one-point solution concepts in cooperative game theory. In this chapter we discuss different formulations of these values, some of their properties and give axiomatic characterizations of the Shapley value.

3.1 The Shapley Value

The Shapley value (cf. [102]) associates to each game $v \in G^N$ one payoff vector in $\mathbb{R}^n$. For a very extensive and interesting discussion on this value the reader is referred to [93].

The first formulation of the Shapley value uses the marginal vectors (see Definition 1.8) of a cooperative TU-game.

Definition 3.1. *The* ***Shapley value*** $\Phi(v)$ *of a game* $v \in G^N$ *is the average of the marginal vectors of the game, i.e.,*

$$\Phi(v) := \frac{1}{n!} \sum_{\sigma \in \pi(N)} m^{\sigma}(v). \tag{3.1}$$

With the aid of (3.1) one can provide a probabilistic interpretation of the Shapley value as follows. Suppose we draw from an urn, containing the elements of $\pi(N)$, a permutation σ (with probability $\frac{1}{n!}$). Then we let the players enter a room one by one in the order σ and give each player the marginal contribution created by him. Then, for each $i \in N$, the i-th coordinate $\Phi_i(v)$ of $\Phi(v)$ is the expected payoff of player i according to this random procedure.

By using Definition 1.8 one can rewrite (3.1) obtaining

$$\Phi_i(v) = \frac{1}{n!} \sum_{\sigma \in \pi(N)} \left(v\left(P^{\sigma}(i) \cup \{i\}\right) - v\left(P^{\sigma}(i)\right)\right). \tag{3.2}$$

Example 3.2. Let $N = \{1, 2, 3\}$, $v(1, 2) = -2$, $v(S) = 0$ if $S \neq \{1, 2\}$. Then the Shapley value is the average of the vectors $(0, -2, 2)$, $(0, 0, 0)$, $(-2, 0, 2)$, $(0, 0, 0)$, $(0, 0, 0)$, and $(0, 0, 0)$, i.e.

$$\Phi(v) = \left(-\frac{1}{3}, -\frac{1}{3}, \frac{2}{3}\right).$$

Remark 3.3. The game in Example 3.2 shows that the Shapley value needs not to be individually rational (cf. Definition 1.27(i)); note that $\Phi_1(v) = -\frac{1}{3} < 0 = v(1)$.

The terms after the summation sign in (3.2) are of the form $v(S \cup \{i\}) - v(S)$, where S is a subset of N not containing i. Note that there are exactly $|S|!\,(n - 1 - |S|)!$ orderings for which one has $P^{\sigma}(i) = S$. The first factor $|S|!$ corresponds to the number of orderings of S and the second factor $(n - 1 - |S|)!$ corresponds to the number of orderings of $N \setminus (S \cup \{i\})$. Using this, we can rewrite (3.2) and obtain

$$\Phi_i(v) = \sum_{S: i \notin S} \frac{|S|!\,(n - 1 - |S|)!}{n!} \left(v(S \cup \{i\}) - v(S)\right). \tag{3.3}$$

Note that $\frac{|S|!(n-1-|S|)!}{n!} = \frac{1}{n}\binom{n-1}{|S|}^{-1}$. This gives rise to a second probabilistic interpretation of the Shapley value. Create a subset S with $i \notin S$ in the following way. First, draw at random a number out of the urn consisting of possible sizes $0, \ldots, n - 1$, where each number (i.e. size) has probability $\frac{1}{n}$ to be drawn. If size s is chosen, draw a set out of the urn consisting of subsets of $N \setminus \{i\}$ of size s, where each set has the same probability $\binom{n-1}{s}^{-1}$ to be drawn. If S is drawn, then one pays player i the amount $v(S \cup \{i\}) - v(S)$. Then, obviously, in view of (3.3), the expected payoff for player i in this random procedure is the Shapley value for player i in the game $v \in G^N$.

Example 3.4. (i) For $v \in G^{\{1,2\}}$ we have

$$\Phi_i(v) = v(i) + \frac{v(1, 2) - v(1) - v(2)}{2} \text{ for each } i \in \{1, 2\}.$$

(ii) The Shapley value $\Phi(v)$ for an additive game $v \in G^N$ is equal to $(v(1), \ldots, v(n))$.

(iii) Let u_S be the unanimity game for $S \subset N$ (cf. (1.1)). Then $\Phi(u_S) = \frac{1}{|S|}e^S$.

The Shapley value satisfies some reasonable properties as introduced in Definition 1.27. More precisely

Proposition 3.5. *The Shapley value satisfies additivity, anonymity, the dummy player property, and efficiency.*

Proof. (*Additivity*) This follows from the fact that $m^\sigma(v+w) = m^\sigma(v)+m^\sigma(w)$ for all $v, w \in G^N$.

(*Anonymity*) We divide the proof into two parts.

(a) First we show that

$$\rho^* \left(m^\sigma (v)\right) = m^{\rho\sigma} \left(v^\rho\right) \text{ for all } v \in G^N \text{ and all } \rho, \sigma \in \pi(N).$$

This follows because for all $i \in N$:

$$\begin{aligned} &\left(m^{\rho\sigma} \left(v^\rho\right)\right)_{\rho\sigma(i)} \\ &= v^\rho \left(\rho\sigma(1), \ldots, \rho\sigma(i)\right) - v^\rho \left(\rho\sigma(1), \ldots, \rho\sigma(i-1)\right) \\ &= v \left(\sigma(1), \ldots, \sigma(i)\right) - v \left(\sigma(1), \ldots, \sigma(i-1)\right) \\ &= \left(m^\sigma (v)\right)_{\sigma(i)} = \rho^* \left(m^\sigma (v)\right)_{\rho\sigma(i)} . \end{aligned}$$

(b) Take $v \in G^N$ and $\rho \in \pi(N)$. Then, using (a), the fact that $\rho \to \rho\sigma$ is a surjection on $\pi(N)$ and the linearity of ρ^*, we obtain

$$\begin{aligned} \Phi\left(v^\rho\right) &= \frac{1}{n!} \sum_{\sigma\in\pi(N)} m^\sigma \left(v^\rho\right) = \frac{1}{n!} \sum_{\sigma\in\pi(N)} m^{\rho\sigma} \left(v^\rho\right) \\ &= \frac{1}{n!} \sum_{\sigma\in\pi(N)} \rho^* \left(m^\sigma (v)\right) = \rho^* \left(\frac{1}{n!} \sum_{\sigma\in\pi(N)} m^\sigma (v)\right) \\ &= \rho^* \left(\Phi\left(v\right)\right). \end{aligned}$$

This proves the anonymity of Φ.

(*Dummy player property*) This follows from (3.3) by noting that $\sum_{S:i\notin S} \frac{|S|!(n-1-|S|)!}{n!} = 1$.

(*Efficiency*) Note that Φ is a convex combination of m^σ's and $\sum_{i\in N} m_i^\sigma(v) = v(N)$ for each $\sigma \in \pi(N)$.

By using the properties listed in Proposition 3.5 one can provide an axiomatic characterization of the Shapley value.

Theorem 3.6. *([102]) A solution $f : G^N \to \mathbb{R}^n$ satisfies additivity, anonymity, the dummy player property, and efficiency if and only if it is the Shapley value.*

Proof. In view of Proposition 3.5 we have only to show that if f satisfies the four properties, then $f = \Phi$.

Take $v \in G^N$. Then $v = \sum_{T \in 2^N \setminus \{\emptyset\}} c_T u_T$ with u_T being the unanimity game for coalition $T \in 2^N \setminus \{\emptyset\}$ and

$$c_T = \sum_{S:S \subset T} (-1)^{|T|-|S|} v(S)$$

(cf. (1.2)). Then by additivity we have $f(v) = \sum_{T \in 2^N \setminus \{\emptyset\}} f(c_T u_T)$, $\Phi(v) = \sum_{T \in 2^N \setminus \{\emptyset\}} \Phi(c_T u_T)$. So we have only to show that for all $T \in 2^N \setminus \{\emptyset\}$ and $c \in \mathbb{R}$:

$$f(cu_T) = \Phi(cu_T). \tag{3.4}$$

Take $T \in 2^N \setminus \{\emptyset\}$ and $c \in \mathbb{R}$. Note first that for all $i \in N \setminus T$:

$$cu_T(S \cup \{i\}) - cu_T(S) = 0 = cu_T(i) \text{ for all } S \in 2^N \setminus \{\emptyset\}.$$

So, by the dummy player property, we have

$$f_i(cu_T) = \Phi_i(cu_T) = 0 \text{ for all } i \in N \setminus T. \tag{3.5}$$

Now, suppose that $i, j \in T$, $i \neq j$. Then there is a $\sigma \in \pi(N)$ with $\sigma(i) = j$, $\sigma(j) = i$, $\sigma(k) = k$ for $k = N \setminus \{i, j\}$. It easily follows that $cu_T = \sigma(cu_T)$. Then anonymity implies that $\Phi(cu_T) = \Phi(\sigma(cu_T)) = \sigma^* \Phi(cu_T)$, $\Phi_{\sigma(i)}(cu_T) = \Phi_i(cu_T)$. So

$$\Phi_i(cu_T) = \Phi_j(cu_T) \text{ for all } i, j \in T, \tag{3.6}$$

and similarly $f_i(cu_T) = f_j(cu_T)$ for all $i, j \in T$.

Then efficiency, (3.5) and (3.6) imply that

$$f_i(cu_T) = \Phi_i(cu_T) = \frac{c}{|T|} \text{ for all } i \in T. \tag{3.7}$$

Now, (3.5) and (3.7) imply (3.4). So, $f(v) = \Phi(v)$ for all $v \in G^N$.

For other axiomatic characterizations of the Shapley value the reader is referred to [56], [74], and [128].

An alternative formula for the Shapley value is in terms of dividends (cf. [55]). The *dividends* d_T for each nonempty coalition T in a game $v \in G^N$ are defined in a recursive manner as follows:

$$d_T(v) := v(T) \text{ for all } T \text{ with } |T| = 1,$$
$$d_T(v) := \frac{v(T) - \sum_{S \subset T, S \neq T} |S| \, d_S(v)}{|T|} \text{ if } |T| > 1.$$

The relation between dividends and the Shapley value is described in the next theorem. It turns out that the Shapley value of a player in a game is the sum of all dividends of coalitions to which the player belongs.

Theorem 3.7. *Let* $v \in G^N$ *and* $v = \sum_{T \in 2^N \setminus \{\emptyset\}} c_T u_T$. *Then*
(i) $|T| \, d_T(v) = c_T$ *for all* $T \in 2^N \setminus \{\emptyset\}$.
(ii) $\Phi_i(v) = \sum_{T: i \in T} d_T(v)$ *for all* $i \in N$.

Proof. We have seen in the proof of Theorem 3.6 that $\Phi(c_T u_T) = \frac{c_T}{|T|} e^T$ for each $T \in 2^N \setminus \{\emptyset\}$, so by additivity,

$$\Phi(v) = \sum_{T \in 2^N \setminus \{\emptyset\}} \frac{c_T}{|T|} e^T.$$

Hence, $\Phi_i(v) = \sum_{T: i \in T} \frac{c_T}{|T|}$. The only thing we have to show is that

$$\frac{c_T}{|T|} = d_T \text{ for all } T \in 2^N \setminus \{\emptyset\}. \tag{3.8}$$

We prove this by induction. If $|T| = 1$, then $c_T = v(T) = d_T(v)$. Suppose (3.8) holds for all $S \subset T$, $S \neq T$. Then $|T| \, d_T(v) = v(T) - \sum_{S \subset T, S \neq T} |S| \, d_S(v) = v(T) - \sum_{S \subset T, S \neq T} c_S = c_T$ because $v(T) = \sum_{S \subset T} c_S$.

Now, we turn to the description of the Shapley value by means of the *multilinear extension* of a game (cf. [83] and [84]).

Let $v \in G^N$. Consider the function $f : [0,1]^n \to \mathbb{R}$ on the hypercube $[0,1]^n$ defined by

$$f(x_1, \ldots, x_n) = \sum_{S \in 2^N} \left(\prod_{i \in S} x_i \prod_{i \in N \setminus S} (1 - x_i) \right) v(S). \tag{3.9}$$

In view of Theorem 1.33, the set of extreme points of $[0,1]^n$ is equal to $\{e^S \mid S \in 2^N\}$.

Proposition 3.8. *Let $v \in G^N$ and f be as above. Then $f\left(e^S\right) = v(S)$ for each $S \in 2^N$.*

Proof. Note that $\prod_{i\in S}\left(e^T\right)^i \prod_{i\in N\setminus S}\left(1-\left(e^T\right)^i\right) = 1$ if $S=T$ and the product is equal to 0 otherwise. Then by (3.9) we have

$$f\left(e^T\right) = \sum_{S\in 2^N}\left(\prod_{i\in S}\left(e^T\right)^i \prod_{i\in N\setminus S}\left(1-\left(e^T\right)^i\right)\right) v(S) = v(T).$$

One can give a probabilistic interpretation of $f(x)$. Suppose that each player $i \in N$, independently, decides whether to cooperate (with probability x_i) or not (with probability $1-x_i$). Then with probability $\prod_{i\in S} x_i \prod_{i\in N\setminus S}(1-x_i)$ the coalition S forms, which has worth $v(S)$. Consequentially, $f(x)$ as given in (3.9) can be seen as the expectation of the worth of the formed coalition.

We denote by $D_k f(x)$ the derivative of f with respect to the k-th coordinate of x. Then we have the following result, describing the Shapley value $\Phi_k(v)$ of a game $v \in G^N$ as the integral along the main diagonal of $[0,1]^n$ of $D_k f$.

Theorem 3.9. *([83]) Let $v \in G^N$ and f be defined as in (3.9). Then $\Phi_k(v) = \int_0^1 (D_k f)(t,\ldots,t)\,dt$ for each $k \in N$.*

Proof. Note that

$$\begin{aligned} &D_k f(x) \\ &= \sum_{T:k\in T}\left(\prod_{i\in T\setminus\{k\}} x_i \prod_{i\in N\setminus T}(1-x_i)\right) v(T) \\ &\quad - \sum_{S:k\notin S}\left(\prod_{i\in S} x_i \prod_{i\in N\setminus(S\cup\{k\})}(1-x_i)\right) v(S) \\ &= \sum_{S:k\notin S}\left(\prod_{i\in S} x_i \prod_{i\in N\setminus(S\cup\{k\})}(1-x_i)\right)\left(v\left(S\cup\{k\}\right) - v(S)\right). \end{aligned}$$

Hence,

$$\begin{aligned} &\int_0^1 (D_k f)(t,t,\ldots,t)\,dt \\ &= \sum_{S:k\notin S}\left(\int_0^1 t^{|S|}(1-t)^{n-|S|-1}\,dt\right)\left(v\left(S\cup\{k\}\right) - v(S)\right). \end{aligned}$$

Using the well known (beta)-integral formula

$$\int_0^1 t^{|S|}(1-t)^{n-|S|-1}\,dt = \frac{|S|!\,(n-1-|S|)!}{n!}$$

we obtain by (3.3)

$$\begin{aligned}&\int_0^1 (D_k f)(t,t,\ldots,t)\,dt\\ &= \sum_{S:k\notin S} \frac{|S|!\,(n-1-|S|)!}{n!}\left(v\left(S\cup\{k\}\right)-v(S)\right)\\ &= \Phi_k(v).\end{aligned}$$

Example 3.10. Let $v \in G^{\{1,2,3\}}$ with $v(1) = v(2) = v(1,2) = 0$, $v(1,3) = 1$, $v(2,3) = 2$, $v(N) = 4$. Then $f(x_1,x_2,x_3) = x_1(1-x_2)x_3 + 2(1-x_1)x_2x_3 + 4x_1x_2x_3 = x_1x_3 + 2x_2x_3 + x_1x_2x_3$ for all $x_1,x_2,x_3 \in [0,1]$. So $D_1 f(x) = x_3 + x_2x_3$. By Theorem 3.9 we obtain

$$\Phi_1(v) = \int_0^1 D_1 f(t,t,t)\,dt = \int_0^1 \left(t+t^2\right)dt = \frac{5}{6}.$$

3.2 The τ-value

The τ-value was introduced in [108] and it is defined for each *quasi-balanced game.* This value is based on the upper vector $M(N,v)$ and the lower vector $m(v)$ of a game $v \in G^N$ (cf. Section 2.2).

Definition 3.11. *A game $v \in G^N$ is called* ***quasi-balanced*** *if*
(i) $m(v) \le M(N,v)$ and
(ii) $\sum_{i=1}^n m_i(v) \le v(N) \le \sum_{i=1}^n M_i(N,v)$.

The set of $|N|$-person quasi-balanced games will be denoted by Q^N.

Proposition 3.12. *If $v \in G^N$ is balanced, then $v \in Q^N$.*

Proof. Let $v \in G^N$ be balanced. Then, by Theorem 2.4, it has a non-empty core.

Let $x \in C(v)$. By Theorem 2.17 we have $m(v) \le x \le M(N,v)$. From this it follows $m(v) \le M(N,v)$ and

$$\sum_{i=1}^n m_i(v) \le \left(\sum_{i=1}^n x_i =\right) v(N) \le \sum_{i=1}^n M_i(N,v).$$

Hence, $v \in Q^N$.

Definition 3.13. *For a game $v \in Q^N$ the **τ-value** $\tau(v)$ is defined by*

$$\tau(v) := \alpha m(v) + (1-\alpha)\, M(N,v)$$

where $\alpha \in [0,1]$ is uniquely determined by $\sum_{i \in N} \tau_i(v) = v(N)$.

Example 3.14. Let $v \in G^{\{1,2,3\}}$ with $v\,(N) = 5$, $v(i) = 0$ for all $i \in N$, $v\,(1,2) = v\,(1,3) = 2$, and $v\,(2,3) = 3$. Then $M(N,v) = (2,3,3)$, $m_1(v) = \max\{0,-1,-1,-1\} = 0$, $m_2(v) = m_3(v) = \max\{0,0,0,0\} = 0$. So $m(v) = 0$ and $v \in Q^{\{1,2,3\}}$. Hence, $\tau(v) = \alpha m(v) + (1-\alpha)\, M(N,v)$ $= \frac{5}{8}\,(2,3,3) = \frac{5}{8} M(N,v)$.

Proposition 3.15. *Let $v \in Q^{\{1,2\}}$. Then*
(i) $C(v) = I(v)$,
(ii) $\tau(v) = \Phi(v)$,
(ii) $\tau(v)$ is in the middle of the core $C(v)$.

Proof. (i) For the lower and upper vectors we have

$$\begin{aligned} m_1(v) &= \max\{v\,(1)\,, v\,(1,2) - M_2(N,v)\} \\ &= \max\{v\,(1)\,, v\,(1,2) - (v\,(1,2) - v\,(1))\} \\ &= v\,(1)\,, \\ M_1(N,v) &= v\,(1,2) - v\,(2)\,. \end{aligned}$$

From $v \in Q^{\{1,2\}}$ it follows $v\,(1) = m_1(v) \leq M_1(N,v) = v\,(1,2) - v\,(2)$, i.e. v is superadditive and its imputation set $I(v)$ is non-empty. Then

$$C(v) = \left\{ x \in I(v) \mid \sum_{i \in S} x_i \geq v(S) \text{ for each } S \subset N \right\} = I(v).$$

(ii) For the Shapley value and for the τ-value we have $\Phi\,(v) = (\Phi_i\,(v))_{i \in \{1,2\}}$ with $\Phi_i\,(v) = \frac{1}{2} v\,(i) + \frac{1}{2}\,(v\,(1,2) - v\,(3-i))$, and
$\tau\,(v) = \frac{1}{2}\,(M(N,v) + m(v))$
$= \frac{1}{2}((v\,(1,2) - v\,(2)\,, v\,(1,2) - v\,(1)) + v\,(1)\,, v\,(2))$
$= \Phi\,(v)\,.$

(iii) From (ii) it follows that $\Phi\,(v) = \tau(v) = \frac{1}{2}\,(f^1 + f^2)$ (cf. (2.1)), which is in the middle of the core $C(v)$.

Example 3.16. Let v be the 99-person game with
$v(N) = 1$, $v(S) = \frac{1}{2}$ if $\{1,2\} \subset S \neq N$,
$v(2,3,4,\ldots,99) = v\,(1,3,4,\ldots,99) = \frac{1}{4}$, and $v(S) = 0$ otherwise.
For the upper and lower vectors we have

$$M(N,v) = \left(\frac{3}{4}, \frac{3}{4}, \frac{1}{2}, \ldots, \frac{1}{2}\right)$$

and

$$m(v) = (0, \ldots, 0)\,.$$

So, $\tau(v) = (1-\alpha)\,M(N,v)$ with $1-\alpha = \frac{4}{200}$. Hence,

$$\tau(v) = \frac{4}{200}\left(\frac{3}{4}, \frac{3}{4}, \frac{2}{4}, \ldots, \frac{2}{4}\right) = \frac{1}{200}\,(3,3,2,\ldots,2)\,.$$

Remark 3.17. The game in Example 3.16 shows that the τ-value may not be in the core $C(v)$ of a game: note that $\tau_1(v) + \tau_2(v) = \frac{6}{200} < \frac{1}{2} = v\,(1,2)$.

Remark 3.18. For an axiomatic characterization of the τ-value the reader is referred to [109].

3.3 The Average Lexicographic Value

In this section we concentrate on balanced cooperative games, which are games with a non-empty core (cf. Definition 1.19 and Theorem 2.4), and introduce for such games the average lexicographic value or AL-value (cf. [111]). Just as in the definition of the Shapley value (cf. Definition 3.1) an averaging of $n!$ vectors takes place which correspond to the $n!$ possible orders of the players in an n-person game. For the Shapley value the vectors are the marginal vectors of the game, while for the AL-value the vectors are the lexicographically optimal points in the core.

Given a game $v \in G^N$ that is balanced and an ordering $\sigma = (\sigma(1), \ldots, \sigma(n))$ of the players in N, the lexicographic maximum of the core $C(v)$ of v with respect to σ is denoted by $L^\sigma(v)$. It is the unique point in $C(v)$ with

$$\begin{aligned}
&(L^\sigma(v))_{\sigma(1)} = \max\{x_{\sigma(1)} \mid x \in C(v)\},\\
&(L^\sigma(v))_{\sigma(2)} = \max\{x_{\sigma(2)} \mid x \in C(v) \text{ with } x_{\sigma(1)} = (L^\sigma(v))_{\sigma(1)}\},\\
&\ldots\\
&(L^\sigma(v))_{\sigma(n)} = \max\{x_{\sigma(n)} \mid x \in C(v)\\
&\qquad\qquad\qquad \text{with } x_{\sigma(i)} = (L^\sigma(v))_{\sigma(i)}, i = 1, \ldots, n-1\}.
\end{aligned}$$

Note that for each $\sigma \in \pi(N)$ the vector $L^\sigma(v)$ is an extreme point of $C(v)$.

Definition 3.19. *The* ***average lexicographic value*** $AL(v)$ *of a balanced game* $v \in G^N$ *is the average of all lexicographically maximal vectors of the core of the game, i.e.,*

$$AL(v) = \frac{1}{n!} \sum_{\sigma \in \pi(N)} L^{\sigma}(v).$$

Example 3.20. Let $v \in G^{\{1,2\}}$ be balanced. Then $v(1,2) \geq v(1) + v(2)$ and $C(v) = co(\{f^1, f^2\})$ with $f^1 = (v(N) - v(2), v(2))$ and $f^2 = (v(1), v(N) - v(1))$. Further $\pi(N) = \{(1,2),(2,1)\}$ and $L^{(1,2)}(v) = f^1$, $L^{(2,1)}(v) = f^2$. So, $AL(v) = \frac{1}{2}(f^1 + f^2) = \big(v(1) + \frac{1}{2}(v(1,2) - v(1) - v(2)),\ v(2) + \frac{1}{2}(v(1,2) - v(1) - v(2))\big)$, the standard solution for 2-person games.

The next two theorems give properties of the AL-value for two special classes of balanced games, namely simplex games and dual simplex games. A game $v \in G^N$ is a *balanced simplex game* (cf. [25] and [26]) if its core $C(v)$ equals its non-empty imputation set $I(v)$. A game $v \in G^N$ is a *balanced dual simplex game* (cf. [25] and [26]) if its core $C(v)$ is equal to its non-empty dual imputation set $I_d(v)$, where $I_d(v) = co(\{g^1(v), \ldots, g^n(v)\})$ with

$$(g^k(v))_i = \begin{cases} v^*(k) = v(N) - v(N\backslash\{k\}) & \text{if } i \neq k, \\ v(N) - \sum_{i \in N\backslash\{k\}} v^*(i) & \text{otherwise.} \end{cases}$$

As it turns out, on these classes of games, the AL-value is closely related with two other single valued solutions concepts: the center of the imputation set (CIS) and the equal split of nonseparable rewards ($ESNR$). Let I^N be the set of all games with non-empty imputation set and I_d^N be the set of all games with non-empty dual imputation set. Then, $CIS : I^N \to \mathbb{R}^n$ is defined by $CIS(v) = \frac{1}{n}\sum_{k=1}^{n} f^k(v)$ and $ESNR : I_d^N \to \mathbb{R}^n$ is defined by $ESNR(v) = \frac{1}{n}\sum_{k=1}^{n} g^k(v)$.

Theorem 3.21. *Let* $v \in G^N$ *be a balanced simplex game. Then* $AL(v) = CIS(v)$.

Proof. Note that $I(v) = co\{f^1(v), \ldots, f^n(v)\}$ and $CIS(v) = \frac{1}{n}\sum_{k=1}^{n} f^k(v)$ where $(f^k(v))_i = v(i)$ for $i \in N\backslash\{k\}$ and $(f^k(v))_k = v(N) - \sum_{i \in N\backslash\{k\}} v(i)$.

Because $L^\sigma(v) = f^{\sigma(1)}(v)$ for each $\sigma \in \pi(N)$ we obtain $AL(v) = \frac{1}{n!} \sum_{\sigma\in\pi(N)} f^{\sigma(1)}(v) = \frac{1}{n} \sum_{k=1}^{n} f^k(v) = CIS(v)$.

For dual simplex games (also called 1-convex games) (cf. [26] and [43]) we give without proof the following results.

Theorem 3.22. *Let $v \in G^N$ be a balanced dual simplex game. Then $AL(v) = ESNR(v)$. Moreover, $AL(v) = \tau(v)$.*

It is easy to see that the AL-value satisfies the following properties: individual rationality, efficiency, core selection, S-equivalence, and symmetry. Also the dummy player property holds for the AL-value because $AL(v)$ is an element of the core for each balanced game $v \in G^N$ and for each $x \in C(v)$,

$$v(i) \leq x_i = \sum_{k=1}^{n} x_k - \sum_{k\in N\setminus\{i\}} x_k \leq v(N) - v(N \setminus \{i\}).$$

So, if $i \in N$ is a dummy player, then $x_i = v(i)$ for each core element and, in particular, $AL_i(v) = v(i)$. Other properties of the average lexicographic value will be discussed in Section 5.2.4.

4

Egalitarianism-based Solution Concepts

4.1 Overview

The principle of egalitarianism is related to the notion of equal share. A completely equal division of $v(N)$ among the players in a game v has little chance to be accepted by all players because it ignores the claims over $v(N)$ of all coalitions $S \in 2^N \setminus \{\emptyset, N\}$. Communities that believe in the egalitarian principle will rather accept an allocation that divides the value of the grand coalition as equally as possible. Such allocations might result using the comparison of coalitions' worth, equal share within groups with high worth being highly desirable. The average worth $a(S,v)$ of $S \in 2^N \setminus \{\emptyset\}$ with respect to v, defined by $a(S,v) := \frac{v(S)}{|S|}$ and called also the per capita value, has played a key role in defining several egalitarian solution concepts. We mention here the equal division core (cf. Definition 2.14), the equal split-off set (cf. Section 4.2), and the constrained egalitarian solution (cf. Subsection 5.2.3).

An alternative way to introduce egalitarian solution concepts is to use egalitarian criteria for comparing income distributions like the Lorenz criterion (cf. [99]) and the Rawlsian criterion (cf. [92]). These criteria use specific binary relations to compare income distributions in order to select one that maximizes the welfare of the worst off players. More precisely, consider a society of n individuals with aggregate income fixed at I units. For any $x \in \mathbb{R}^n_+$ denote by $\widehat{x} = (\widehat{x}_1, \ldots, \widehat{x}_n)$ the vector obtained by rearranging its coordinates in a non-decreasing order, that is, $\widehat{x}_1 \leq \widehat{x}_2 \leq \ldots \leq \widehat{x}_n$. For any $x, y \in \mathbb{R}^n_+$ with $\sum_{i=1}^n x_i = \sum_{i=1}^n y_i = I$, we say that x *Lorenz dominates* y, and denote it by $x \succ_L y$, iff $\sum_{i=1}^p \widehat{x}_i \geq \sum_{i=1}^p \widehat{y}_i$ for all $p \in \{1, \ldots, n-1\}$, with at least one strict inequality. The Rawlsian criterion is based on a leximin domination. We say that x *leximin dominates* y, denoted by $x \succ_{lex} y$, if there exists $k \in \{0, 1, \ldots, n-1\}$ such that $\hat{x}_i = \hat{y}_i$ for $i = 1, \ldots, k$ and $\hat{x}_{k+1} > \hat{y}_{k+1}$.

In most of the game theory literature about egalitarianism it is agreed upon the fact that in order to call an allocation egalitarian it should be necessarily maximal according to the Lorenz criterion. For arbitrary TU games existing egalitarianism-based solution concepts using the Lorenz domination with respect to distributions of $v(N)$ are the constrained egalitarian solution (cf. [46]), the strong-constrained egalitarian allocations (cf. [47]), the egalitarian set, the pre-egalitarian set and the stable egalitarian set (cf. [3]). For balanced games the Lorenz criterion plays a central role for the Lorenz solution (cf. [59]), the Lorenz stable set and the egalitarian core (cf. [2]).

It is also interesting to look at solution concepts which are not qualified as egalitarian and disclose whether they apply some egalitarian treatment to players or coalitions. We refer the reader interested in egalitarian ideas behind the notions of the prenucleolus (cf. [96]), the λ-prenucleolus (cf. [94]), the nucleolus (cf. [96]) and the Shapley value (cf. [102]) to [1].

4.2 The Equal Split-Off Set

We introduce now a new set valued solution concept for cooperative games with transferable utility that we call the equal split-off set (cf. [21]). More precisely, we consider a world N of n players, $N = \{1, \ldots, n\}$, who believe in equal share cooperation and play the game $v \in G^N$. We assume that the entire set of players will cooperate and deal with the question how the whole amount of money $v(N)$ generated by N should be divided among the players by considering the following step-wise process.

First, one of the coalitions with maximal average worth, say T_1, forms and the players in T_1 divide equally the worth $v(T_1)$. In step 2 one of the coalitions in $N \setminus T_1$ with maximal average marginal worth w.r.t. T_1, say T_2, forms, joins costless T_1, and divides equally the increase in value $v(T_2 \cup T_1) - v(T_1)$ among its members. The process stops when a partition of N of the form $\langle T_1, \ldots, T_K \rangle$ for some $1 \leq K \leq n$ is reached. This procedure generates an efficient payoff vector $x \in \mathbb{R}^n$ which we call an equal split-off allocation. The equal split-off set is then defined as consisting of all equal split-off allocations.

4.2.1 The Equal Split-Off Set for General Games

Let $v \in G^N$ and $\pi = \langle T_1, \ldots, T_K \rangle$ be an ordered partition of the player set N. We set $v_1 := v$, and for each $k \in \{2, \ldots, K\}$ we define the *marginal game*

$$v_k : 2^{N \setminus \left(\cup_{s=1}^{k-1} T_s\right)} \longrightarrow \mathbb{R}$$

by

$$\begin{aligned} v_k(S) &:= v_{k-1}\left(T_{k-1} \cup S\right) - v_{k-1}\left(T_{k-1}\right) \\ &= v\left(\left(\cup_{s=1}^{k-1} T_s\right) \cup S\right) - v\left(\cup_{s=1}^{k-1} T_s\right). \end{aligned} \tag{4.1}$$

We call the partition $\pi = \langle T_1, \ldots, T_K \rangle$ of N a *suitable ordered partition with respect to the game* $v \in G^N$ if

$$T_k \in \arg \max_{S \in 2^{N \setminus \left(\cup_{s=1}^{k-1} T_s\right)} \setminus \{\emptyset\}} \frac{v_k(S)}{|S|}$$

for all $k \in \{1, \ldots, K\}$.

Given such a partition π, the *equal split-off allocation for* v *generated by* π is the efficient payoff vector $x = (x_i)_{i \in N} \in \mathbb{R}^n$, where for all $T_k \in \pi$ and all $i \in T_k$, $x_i = \frac{v_k(T_k)}{|T_k|}$.

Definition 4.1. *The* **equal split-off set** $ESOS(v)$ *of a game* $v \in G^N$ *is the set*

$$\{x \in \mathbb{R}^n \mid \exists \pi \text{ s.t. } x \text{ is an equal split-off allocation for } v \text{ generated by } \pi\}.$$

In order to illustrate this solution concept, let us have a look at the following examples:

Example 4.2. (2-person superadditive games) Let v be a game on the player set $N = \{1, 2\}$ satisfying $v(1,2) \geq v(1) + v(2)$. Suppose without loss of generality that $v(1) \geq v(2)$ and consider the following four cases:

(i) $v(1) > \frac{1}{2} v(1,2)$. Then $\langle \{1\}, \{2\} \rangle$ is the unique suitable ordered partition and

$$ESOS(v) = \{(v(1), v(1,2) - v(1))\};$$

(ii) $v(2) < v(1) = \frac{1}{2} v(1,2)$. In this case

$$ESOS(v) = \left\{ \left(\frac{1}{2} v(1,2), \frac{1}{2} v(1,2) \right) \right\}$$

corresponding to the suitable ordered partitions $\langle \{1\}, \{2\} \rangle$ and $\langle \{1, 2\} \rangle$;

(iii) $v(2) = v(1) = \frac{1}{2}v(1,2)$. Also here

$$ESOS(v) = \left\{\left(\frac{1}{2}v(1,2), \frac{1}{2}v(1,2)\right)\right\} = \{(v(1), v(2))\},$$

corresponding to the three suitable ordered partitions $\langle\{1\},\{2\}\rangle$, $\langle\{2\},\{1\}\rangle$, and $\langle\{1,2\}\rangle$;

(iv) $v(1) < \frac{1}{2}v(1,2)$. Then $\langle\{1,2\}\rangle$ is the unique suitable ordered partition and

$$ESOS(v) = \left\{\left(\frac{1}{2}v(1,2), \frac{1}{2}v(1,2)\right)\right\}.$$

Example 4.3. (Simple games) Given a simple game v on player set N (cf. Definition 1.3), we denote the set of all minimal winning coalitions (cf. Definition 1.5) with a smallest cardinality by W^s. In this case we have $ESOS(v) = \left\{\frac{1}{|S|}e^S \mid S \in W^s\right\}$ because for any suitable ordered partition $\langle T_1, \ldots, T_K\rangle$ we will have $T_1 \in W^s$, and all players in T_1 will receive $\frac{1}{|T_1|}$ whereas the players in $N \setminus T_1$ will receive payoff 0.

Example 4.4. (Glove games) Let v be a glove game on player set N (see Example 1.2). If $|L| = |R|$, then $ESOS(v) = \left\{\left(\frac{1}{2}, \ldots, \frac{1}{2}\right)\right\}$ that can be generated by many suitable ordered partitions, where each element T_k of such a partition has the property that $|T_k \cap L| = |T_k \cap R|$. In case $|L| > |R|$ each element $x \in ESOS(v)$ satisfies $x_i = \frac{1}{2}$ for each $i \in R$ and for $|R|$ elements of L, and $x_i = 0$ for the other elements of L. Conversely, all elements of this type belong to $ESOS(v)$.

Example 4.5. (A 2-person non-superadditive game) Let v be a game on the player set $N = \{1,2\}$ satisfying $v(\emptyset) = v(1,2) = 0$, $v(1) = 3$, and $v(2) = 2$. Then $\langle\{1\},\{2\}\rangle$ is the unique suitable ordered partition and $ESOS(v) = \{(3,-3)\}$.

One can easily check that the suitable ordered partitions generating equal split-off allocations in the above examples satisfy a monotonicity property w.r.t. average worth as stated in

Proposition 4.6. *Let $v \in G^N$ and $\langle T_1, \ldots, T_K\rangle$ be a suitable ordered partition of N w.r.t. v. Then*

$$\max_{S \in 2^{N \setminus \left(\cup_{s=1}^{k-1} T_s\right)} \setminus \{\emptyset\}} \frac{v_k(S)}{|S|} \geq \max_{S \in 2^{N \setminus \left(\cup_{s=1}^{k} T_s\right)} \setminus \{\emptyset\}} \frac{v_{k+1}(S)}{|S|}$$

for all $k \in \{1, \ldots, K-1\}$.

Proof. It follows from the definition of T_k that

$$\frac{v\left(\cup_{s=1}^{k} T_s\right) - v\left(\cup_{s=1}^{k-1} T_s\right)}{|T_k|} \geq \frac{v\left(\cup_{s=1}^{k+1} T_s\right) - v\left(\cup_{s=1}^{k-1} T_s\right)}{|T_k| + |T_{k+1}|}.$$

By adding and subtracting $v(\cup_{s=1}^{k} T_s)$ in the numerator of the right-hand term, we obtain

$$\frac{v\left(\cup_{s=1}^{k} T_s\right) - v\left(\cup_{s=1}^{k-1} T_s\right)}{|T_k|} \geq$$
$$\frac{v\left(\cup_{s=1}^{k+1} T_s\right) - v(\cup_{s=1}^{k} T_s) + v(\cup_{s=1}^{k} T_s) - v\left(\cup_{s=1}^{k-1} T_s\right)}{|T_k| + |T_{k+1}|}.$$

This inequality is equivalent to

$$\left(v\left(\cup_{s=1}^{k} T_s\right) - v\left(\cup_{s=1}^{k-1} T_s\right)\right)|T_k| + \left(v\left(\cup_{s=1}^{k} T_s\right) - v\left(\cup_{s=1}^{k-1} T_s\right)\right)|T_{k+1}|$$
$$\geq \left(v\left(\cup_{s=1}^{k+1} T_s\right) - v\left(\cup_{s=1}^{k} T_s\right)\right)|T_k| + \left(v\left(\cup_{s=1}^{k} T_s\right) - v\left(\cup_{s=1}^{k-1} T_s\right)\right)|T_k|,$$

which is at its turn equivalent to

$$\left(v\left(\cup_{s=1}^{k} T_s\right) - v\left(\cup_{s=1}^{k-1} T_s\right)\right)|T_{k+1}| \geq \left(v\left(\cup_{s=1}^{k+1} T_s\right) - v\left(\cup_{s=1}^{k} T_s\right)\right)|T_k|.$$

We refer the reader to [127] for an axiomatic characterization of the equal split-off set on the class of arbitrary crisp games that uses a consistency property à la Hart-Mas-Colell (cf. [56]).

4.2.2 The Equal Split-Off Set for Superadditive Games

We consider now an interesting additional property of our solution concept on the class of superadditive games (cf. Definition 1.14). As it turns out, the equal split-off set of a superadditive game is a refinement of the equal division core of that game (cf. Definition 2.14).

Theorem 4.7. *Let v be a superadditive game. Then $ESOS(v) \subset EDC(v)$.*

Proof. Let $x \in ESOS(v)$ be generated by the suitable ordered partition $\langle T_1, \dots, T_K \rangle$. Take $S \in 2^N \setminus \{\emptyset\}$. We have to prove that there is $i \in S$ such that $x_i \geq \frac{v(S)}{|S|}$.

Let $m \in \{1, \dots, K\}$ be the smallest number such that $T_m \cap S \neq \emptyset$. Then

$$\begin{aligned}\frac{v(S)}{|S|} &\leq \frac{v\left(\left(\cup_{s=1}^{m-1} T_s\right) \cup S\right) - v\left(\cup_{s=1}^{m-1} T_s\right)}{|S|} \\ &\leq \frac{v\left(\cup_{s=1}^{m} T_s\right) - v\left(\cup_{s=1}^{m-1} T_s\right)}{|T_m|} \\ &= \frac{v_m(T_m)}{|T_m|} = \max_{T \in 2^{N \setminus \left(\cup_{s=1}^{m-1} T_s\right)} \setminus \{\emptyset\}} \frac{v_m(T)}{|T|},\end{aligned}$$

where the first inequality follows from the superadditivity of v and the second inequality from the definition of T_m. Note that

$$x_i = \max_{T \in 2^{N \setminus \left(\cup_{s=1}^{m-1} T_s\right)} \setminus \{\emptyset\}} \frac{v_m(T)}{|T|} \geq \frac{v(S)}{|S|}$$

for each $i \in T_m \cap S$. So, $x \in EDC(v)$ implying that $ESOS(v) \subset EDC(v)$.

The next example provides a game for which the equal split-off set is a strict subset of the equal division core.

Example 4.8. Let $N = \{1, 2, 3\}$ and v be a glove game with $L = \{1, 2\}$ and $R = \{3\}$ (see Example 1.2). Then $EDC(v) = \left\{x \in I(v) \mid x_3 \geq \frac{1}{2}\right\}$ and $ESOS(v) = \left\{\left(\frac{1}{2}, 0, \frac{1}{2}\right), \left(0, \frac{1}{2}, \frac{1}{2}\right)\right\}$.

Remark 4.9. Clearly, by Theorem 4.7, each equal split-off allocation for a superadditive game is individually rational. As illustrated in Example 4.5, this needs not be the case for non-superadditive games.

5

Classes of Cooperative Crisp Games

In this chapter we consider three classes of cooperative crisp games: totally balanced games, convex games, and clan games. We introduce basic characterizations of these games and discuss special properties of the set-valued and one-point solution concepts introduced so far. Moreover, we relate the corresponding games with the concept of a population monotonic allocation scheme as introduced in [107]. We present the notion of a bi-monotonic allocation scheme for total clan games and the constrained egalitarian solution (cf. [45] and [46]) for convex games.

5.1 Totally Balanced Games

5.1.1 Basic Characterizations and Properties of Solution Concepts

Let $v \in G^N$. The game v is called *totally balanced* if all its subgames (cf. Definition 1.9) are balanced (cf. Definition 1.19). Equivalently, the game v is totally balanced if $C(v_T) \neq \emptyset$ for all $T \in 2^N \setminus \{\emptyset\}$ (cf. Theorem 2.4).

Example 5.1. Let $v \in G^{\{1,2,3,4\}}$ with $v(S) = 0, 0, 1, 2$ if $|S| = 0, 1, 3, 4$ respectively, and $v(1,2) = v(1,3) = v(2,3) = 1$, $v(1,4) = v(2,4) = v(3,4) = 0$. Then $\left(\frac{1}{2}, \frac{1}{2}, \frac{1}{2}, \frac{1}{2}\right) \in C(v)$, so v is balanced, but the subgame v_T with $T = \{1,2,3\}$ is not balanced, so the game v is not totally balanced.

Example 5.2. Let $v \in G^{\{1,2,3,4\}}$ with $v(1,2) = v(3,4) = \frac{1}{2}$, $v(1,2,3) = v(2,3,4) = v(1,2,4) = v(1,3,4) = \frac{1}{2}$, $v(1,2,3,4) = 1$, and $v(S) = 0$

for all other $S \in 2^N$. Then $\left(\frac{1}{4}, \frac{1}{4}, \frac{1}{4}, \frac{1}{4}\right) \in C(v)$. Furthermore, $\left(0, \frac{1}{2}, 0\right)$ is an element of the core of the 3-person subgames and it easily follows that also the one- and two-person games have non-empty cores. Hence, the game v is totally balanced.

The following theorem relates totally balanced games that are non-negative (cf. Definition 1.12) and additive games (cf. Definition 1.13).

Theorem 5.3. *([64]) Let $v \in G^N$ be totally balanced and non-negative. Then v is the intersection of $2^n - 1$ additive games.*

Proof. Let $v \in G^N$ be as above. For each $T \in 2^N \setminus \{\emptyset\}$ consider the corresponding subgame v_T and take $x_T \in C(v_T)$. Define $y_T \in \mathbb{R}^n$ by $y_T^i := x_T^i$ if $i \in T$ and $y_T^i := \alpha$ if $i \in N \setminus T$, where $\alpha := \max_{S \in 2^N} v(S)$. We prove that v is equal to $\wedge_{T \in 2^N \setminus \{\emptyset\}} w_T$ $\left(:= \min\left\{w_T \mid T \in 2^N \setminus \{\emptyset\}\right\}\right)$, where w_T is the additive game with $w_T(i) = y_T^i$ for all $i \in N$.

We have to show that for $S \in 2^N \setminus \{\emptyset\}$,

$$\min\left\{w_T(S) \mid T \in 2^N \setminus \{\emptyset\}\right\} = v(S).$$

This follows from

(a) $w_S(S) = \sum_{i \in S} y_S^i = \sum_{i \in S} x_S^i = v_S(S) = v(S)$,
(b) $w_T(S) \geq \alpha \geq v(S)$ if $S \setminus T \neq \emptyset$, where the first inequality follows from the non-negativity of the game,
(c) $w_T(S) = \sum_{i \in S} x_S^i \geq v_T(S) = v(S)$ if $S \subset T$.

Nice examples of totally balanced games are games arising from flow situations with dictatorial control. A flow situation consists of a directed network with two special nodes called the source and the sink. For each arc there are a capacity constraint and a constraint with respect to the allowance to use that arc. Furthermore, with the aid of a simple game (cf. Definition 1.3) for each arc, one can describe which coalitions are allowed to use the arc. These are the coalitions which are winning (cf. Definition 1.4). Such games are called *control games* in this context. The value of a coalition S is the maximal flow per unit of time through the network from source to sink, where one uses only arcs which are controlled by S. Clearly, a dictatorial control game is a control game in which the arcs are controlled by dictators (cf. Definition 1.6).

One can show (cf. [64]) that each flow game with dictatorial control is totally balanced and non-negative. The converse is also true as shown in

Theorem 5.4. *([64]) Let $v \in G^N$ be totally balanced and non-negative. Then v is a flow game with dictatorial control.*

Proof. The minimum $v \wedge w$ of two flow games $v, w \in G^N$ with dictatorial control is again such a flow game: make a series connection of the flow networks of v and w. Also an additive game v is a flow game with dictatorial control. Combining these facts with Theorem 5.3 completes the proof.

5.1.2 Totally Balanced Games and Population Monotonic Allocation Schemes

The class of totally balanced games includes the class of games with a *population monotonic allocation scheme* (pmas). The latter concept was introduced in [107]. The idea here is that because of the complexity of the coalition formation process, players may not necessarily achieve full efficiency (if the game is superadditive it is efficient for the players to form the grand coalition). In order to take the possibility of partial cooperation into account, a pmas specifies not only how to allocate $v(N)$ but also how to allocate the value $v(S)$ of every coalition $S \in 2^N \setminus \{\emptyset\}$. Moreover, it reflects the intuition that there is "strength in numbers": the share allocated to each member is nondecreasing in the coalition size.

Definition 5.5. *Let $v \in G^N$. A scheme $a = (a_{iS})_{i \in S, S \in 2^N \setminus \{\emptyset\}}$ of real numbers is a* **population monotonic allocation scheme** *(pmas) of v if*

(i) $\sum_{i \in S} a_{iS} = v(S)$ for all $S \in 2^N \setminus \{\emptyset\}$,
(ii) $a_{iS} \leq a_{iT}$ for all $S, T \in 2^N \setminus \{\emptyset\}$ with $S \subset T$ and $i \in S$.

Definition 5.6. *Let $v \in G^N$. An imputation $b \in I(v)$ is* **pmas extendable** *if there exist a pmas $a = (a_{iS})_{i \in S, S \in 2^N \setminus \{\emptyset\}}$ such that $a_{iN} = b_i$ for each player $i \in N$.*

As it can be easily derived from Definition 5.5, a necessary condition for a game to posses a pmas is that the game is totally balanced. This condition is also a sufficient one for games with at most three players: one can easily show that every core element of such a game is pmas extendable. However, if the number of players is at least four, the existence of a pmas is not guaranteed as the next example shows (cf. [107]).

Example 5.7. Let $v \in G^{\{1,2,3,4\}}$ with $v(i) = 0$ for $i = 1, \ldots, 4$, $v(1,2) = v(3,4) = 0$, $v(1,3) = v(1,4) = v(2,3) = v(2,4) = 1$, $v(S) = 1$ for all S with $|S| = 3$, and $v(N) = 2$. The core of this game is the line segment joining $(0,0,1,1)$ and $(1,1,0,0)$. One can easily see that each subgame of this game has a nonempty core, i.e. the game is totally balanced. However, the game lacks a pmas as it can be shown by the following argument: every pmas must satisfy $a_{1N} \geq a_{1\{1,3,4\}} = 1$, $a_{2N} \geq a_{2\{2,3,4\}} = 1$, $a_{3N} \geq a_{3\{1,2,3\}} = 1$, and $a_{4N} \geq a_{4\{1,2,4\}} = 1$. Hence, $\sum_{i \in N} a_{iN} \geq 4$, which is not feasible.

For general necessary and sufficient conditions for a game to possess a pmas the reader is referred to [107].

5.2 Convex Games

This class of cooperative games was introduced in [104]. In addition to many equivalent characterizations of convex games, these games have nice properties: the core of such a game is the unique stable set, its extreme points can be easily described and it is additive on the cone of convex games (cf. [41] and [112]). Moreover, the Shapley value coincides with the barycenter of the core in the sense that it is the average of the marginal vectors. It turns out that on the cone of convex games the AL-value and the Shapley value coincide. Furthermore, the equal split-off set of a convex game coincides with the constrained egalitarian solution of the game (cf. Subsection 5.2.3).

5.2.1 Basic Characterizations

Definition 5.8. *A game $v \in G^N$ is called* ***convex*** *iff*

$$v(S \cup T) + v(S \cap T) \geq v(S) + v(T) \text{ for all } S, T \in 2^N. \tag{5.1}$$

Definition 5.9. *A game $v \in G^N$ is called* ***concave*** *iff $-v$ is convex.*

In what follows the set of convex games on player set N will be denoted by CG^N.

In the next theorem we give five characterizations of convex games. Characterizations (ii) and (iii) show that for convex games the gain made when individuals or groups join larger coalitions is higher than when they join smaller coalitions. Characterizations (iv) and (v) deal with the relation between the core and the Weber set (cf. [104], [62], [37], [36]).

Theorem 5.10. *Let $v \in G^N$. The following five assertions are equivalent.*

(i) $v \in CG^N$;
(ii) For all $S_1, S_2, U \in 2^N$ *with* $S_1 \subset S_2 \subset N \setminus U$ *we have*

$$v(S_1 \cup U) - v(S_1) \leq v(S_2 \cup U) - v(S_2); \tag{5.2}$$

(iii) For all $S_1, S_2 \in 2^N$ *and* $i \in N$ *such that* $S_1 \subset S_2 \subset N \setminus \{i\}$ *we have*

$$v(S_1 \cup \{i\}) - v(S_1) \leq v(S_2 \cup \{i\}) - v(S_2); \tag{5.3}$$

(iv) All $n!$ *marginal vectors* $m^\sigma(v)$ *of* v *are elements of the core* $C(v)$ *of* v;
(v) $W(v) = C(v)$.

Proof. We show *(i)* ⇒ *(ii)*, *(ii)* ⇒ *(iii)*, *(iii)* ⇒ *(iv)*, *(iv)* ⇒ *(v)*, *(v)* ⇒ *(i)*.

(a) Suppose that *(i)* holds. Take $S_1, S_2, U \in 2^N$ with $S_1 \subset S_2 \subset N \setminus U$. From (5.1) with $S_1 \cup U$ in the role of S and S_2 in the role of T we obtain (5.2) by noting that $S \cup T = S_2 \cup U$, $S \cap T = S_1$. Hence, *(i)* implies *(ii)*.
(b) That *(ii)* implies *(iii)* is trivial (take $U = \{i\}$).
(c) Suppose that *(iii)* holds. Let $\sigma \in \pi(N)$ and take m^σ. Then $\sum_{k=1}^n m_k^\sigma = v(N)$. To prove that $m^\sigma \in C(v)$ we have to show that for $S \in 2^N : \sum_{k \in S} m_k^\sigma \geq v(S)$.
Let $S = \{\sigma(i_1), \ldots, \sigma(i_k)\}$ with $i_1 < \ldots < i_k$. Then

$$\begin{aligned} & v(S) \\ &= \sum_{r=1}^{k} \left(v\left(\sigma(i_1), \ldots, \sigma(i_r)\right) - v\left(\sigma(i_1), \ldots, \sigma(i_{r-1})\right)\right) \\ &\leq \sum_{r=1}^{k} \left(v\left(\sigma(1), \ldots, \sigma(i_r)\right) - v\left(\sigma(1), \ldots, \sigma(i_r - 1)\right)\right) \\ &= \sum_{r=1}^{k} m_{\sigma(i_r)}^\sigma = \sum_{k \in S} m_k^\sigma, \end{aligned}$$

where the inequality follows from *(iii)* applied to $i := \sigma(i_r)$ and $S_1 := \{\sigma(i_1), \ldots, \sigma(i_{r-1})\} \subset S_2 := \{\sigma(1), \ldots, \sigma(i_r - 1)\}$ for $r \in \{1, \ldots, k\}$. This proves that *(iii)* implies *(iv)*.

(d) Suppose that *(iv)* holds. Since $C(v)$ is a convex set, we have $C(v) \supset co\{m^\sigma \mid \sigma \in \pi(N)\} = W(v)$. From Theorem 2.20 we know that $C(v) \subset W(v)$. Hence, *(v)* follows from *(iv)*.

(e) Finally, we prove that *(v)* implies *(i)*. Take $S, T \in 2^N$. Then, there is $\sigma \in \pi(N)$ and $d, t, u \in \mathbb{N}$ with $0 \leq d \leq t \leq u \leq n$ such that $S \cap T = \{\sigma(i_1), \ldots, \sigma(d)\}$, $T \setminus S = \{\sigma(d+1), \ldots, \sigma(t)\}$, $S \setminus T = \{\sigma(t+1), \ldots, \sigma(u)\}$, $N \setminus (S \cup T) = \{\sigma(u+1), \ldots, \sigma(n)\}$. From *(v)* follows that $m^\sigma \in C(v)$; hence,

$$v(S) \leq \sum_{i \in S} m_i^\sigma. \tag{5.4}$$

We have also that

$$\begin{aligned}
&\sum_{i \in S} m_i^\sigma && (5.5)\\
&= \sum_{r=1}^{d} \left(v(A_r) - v(A_{r-1})\right) + \sum_{k=1}^{u-t} v(T \cup B_{t+k}) - v(T \cup B_{t+k-1})\\
&= v(S \cap T) + v(S \cup T) - v(T),
\end{aligned}$$

where

$$A_r = \{\sigma(1), \ldots, \sigma(r)\},$$
$$A_{r-1} = A_r \setminus \{\sigma(r)\},$$
$$B_{t+k} = \{\sigma(t+1), \ldots, \sigma(t+k)\},$$

and

$$B_{t+k-1} = B_{t+k} \setminus \{\sigma(t+k)\}.$$

Combining (5.4) and (5.5) yields (5.1). This completes the proof.

Definition 5.11. *([97]) A game $v \in G^N$ is called* ***exact*** *if for each $S \in 2^N \setminus \{\emptyset\}$ there is an $x \in C(v)$ with $\sum_{i \in S} x_i = v(S)$.*

Remark 5.12. It is not difficult to see that a convex game is exact.

Remark 5.13. A characterization of convex games using the exactness of its subgames (cf. Definition 1.9) can be found in [15] and [8].

5.2.2 Convex Games and Population Monotonic Allocation Schemes

As we have pointed out in Section 5.1.2, a necessary condition for the existence of a pmas (cf. Definition 5.5) is the total balancedness of the game. A sufficient condition for the existence of a pmas is the convexity of the game. In order to see this we will need the following definitions.

For all $\rho \in \pi(N)$ and all $i \in N$, let

$$N(\rho, i) = \{j \in N \mid \rho(j) \leq \rho(i)\}.$$

One generalizes the definition of a marginal contribution vector (cf. Definition 1.8) as follows.

Definition 5.14. *Let $v \in G^N$ and $\rho \in \pi(N)$. The* ***extended vector of marginal contributions*** *associated with ρ is the vector $a^\rho = \left(a^\rho_{iS}\right)_{i\in S, S\in 2^N\setminus\{\emptyset\}}$ defined component-wise by*

$$a^\rho_{iS} = v\left(N(\rho, i) \cap S\right) - v\left(\left(N(\rho, i) \cap S\right) \setminus \{i\}\right).$$

Proposition 5.15. *([107]) Let $v \in CG^N$. Then every extended vector of marginal contributions is a pmas for v.*

Proof. Take $v \in CG^N$, $\rho \in \pi(N)$, and a^ρ. Pick an arbitrary $S \in 2^N\setminus\{\emptyset\}$ and rank all players $i \in S$ in increasing order of $\rho(i)$. Let $i, i' \in S$ be two players such that i' immediately follows i. Observe that

$$a^\rho_{iS} = v\left(N(\rho, i) \cap S\right) - v\left(\left(N(\rho, i) \cap S\right) \setminus \{i\}\right),$$

and

$$\begin{aligned} a^\rho_{i'S} &= v\left(N(\rho, i') \cap S\right) - v\left(\left(N(\rho, i') \cap S\right) \setminus \{i'\}\right) \\ &= v\left(N(\rho, i') \cap S\right) - v\left(\left(N(\rho, i) \cap S\right)\right). \end{aligned}$$

Therefore, $a^\rho_{iS} + a^\rho_{i'S} = v\left(N(\rho, i') \cap S\right) - v\left(\left(N(\rho, i) \cap S\right) \setminus \{i\}\right)$. Repeating this argument leads to $\sum_{i\in S} a^\rho_{iS} = v(S)$, which establishes the feasibility of a^ρ.

As for the monotonicity property in Definition 5.5, note that if $i \in S \subset T \subset N$, then $S \cap N(\rho, i) \subset T \cap N(\rho, i)$ for all $i \in N$. Hence, by the convexity of v we have $a^\rho_{iS} \leq a^\rho_{iT}$. This completes the proof.

According to [104] the core of a convex game is a polytope whose extreme points are the (usual) marginal contribution vectors (cf. Theorem 5.10). Because every convex combination of pmas of a game v is itself a pmas for that game, one obtains

Proposition 5.16. *Let $v \in CG^N$ and $b = (b_i)_{i \in N} \in C(v)$. Then b is pmas extendable.*

Definition 5.17. *Let $v \in G^N$. The* ***extended Shapley value*** *of v is the vector $\widetilde{\Phi}(v)$ defined component-wise as follows: for all $S \in 2^N \setminus \{\emptyset\}$ and all $i \in S$,*

$$\widetilde{\Phi}_{iS}(v) = \Phi_i(v_S),$$

where $\Phi(v_S) = (\Phi_i(v_S))_{i \in S}$ is the Shapley value of the game v_S.

As shown in [107], the extended Shapley value is the aritmethic average of the extended vectors of marginal contributions (cf. Definition 5.14). Therefore, one obtains

Proposition 5.18. *Let $v \in CG^N$. Then the extended Shapley value of v is a pmas for v.*

5.2.3 The Constrained Egalitarian Solution for Convex Games

Another interesting element of the core of a game $v \in CG^N$ is the constrained egalitarian allocation $E(v)$ introduced in [46] which can be described in a simple way and found easily in a finite number of steps. Two lemmas in which the average worth $\frac{v(S)}{|S|}$ of a nonempty coalition S with respect to the characteristic function v plays a role, are used further.

Lemma 5.19. *Let $v \in CG^N$ and $L(v) := \arg\max_{C \in 2^N \setminus \{\emptyset\}} \frac{v(C)}{|C|}$. Then*

(i) The set $L(v) \cup \{\emptyset\}$ is a lattice, i.e. for all $S_1, S_2 \in L(v) \cup \{\emptyset\}$ we have $S_1 \cap S_2 \in L(v) \cup \{\emptyset\}$ and $S_1 \cup S_2 \in L(v) \cup \{\emptyset\}$;
(ii) In $L(v)$ there is a maximal element with respect to $\subset$ namely

$$\cup \{S \mid S \in L(v)\}.$$

Proof. (i) Let $\alpha := \max_{C \in 2^N \setminus \{\emptyset\}} \frac{v(C)}{|C|}$ and suppose $\frac{v(S_1)}{|S_1|} = \alpha = \frac{v(S_2)}{|S_2|}$ for some $S_1, S_2 \in 2^N \setminus \{\emptyset\}$. We have to prove that

$$\frac{v(S_1 \cup S_2)}{|S_1 \cup S_2|} = \alpha \text{ and } v(S_1 \cap S_2) = \alpha \, |S_1 \cap S_2| . \tag{5.6}$$

We have

$$\begin{aligned}
&v(S_1 \cup S_2) + v\,(S_1 \cap S_2) \\
&= \frac{v(S_1 \cup S_2)}{|S_1 \cup S_2|}\,|S_1 \cup S_2| + \frac{v(S_1 \cap S_2)}{|S_1 \cap S_2|}\,|S_1 \cap S_2| \\
&\leq \alpha\,|S_1 \cup S_2| + \alpha\,|S_1 \cap S_2| = \alpha\,|S_1| + \alpha\,|S_1| \\
&= v(S_1) + v(S_2) \leq v(S_1 \cup S_2) + v(S_1 \cap S_2),
\end{aligned}$$

where the first inequality follows from the definition of α and the second inequality follows from $v \in CG^N$. So everywhere we have equalities, which proves (5.6).

(ii) This assertion follows immediately from (i) and the finiteness of $L(v)$.

Lemma 5.20. *Let $v \in CG^N$ and $S \subset N, S \neq N$. Then $v^{-S} \in CG^{N \setminus S}$, where*

$$v^{-S}\,(T) := v\,(S \cup T) - v(S) \textit{ for all } T \in 2^{N \setminus S}.$$

Proof. Let $T_1 \subset T_2 \subset (N \setminus S) \setminus \{i\}$ where $i \in N \setminus S$. We have to prove that $v^{-S}\,(T_1 \cup \{i\}) - v(T_1) \leq v^{-S}\,(T_2 \cup \{i\}) - v(T_2)$. Notice that this is equivalent to prove that $v\,(S \cup T_1 \cup \{i\}) - v(S \cup T_1) \leq v\,(S \cup T_2 \cup \{i\}) - v(S \cup T_2)$ which follows by the convexity of v.

Remark 5.21. It is easy to see that v^{-S} is a marginal game (cf. (4.1)).

Given these two lemmas, one can find the egalitarian allocation $E\,(v)$ of a game $v \in CG^N$ according to the following algorithm (cf. [46]).

In Step 1 of the algorithm one considers the game $\langle N_1, v_1 \rangle$ with $N_1 := N$, $v_1 := v$, and the per capita value $\frac{v_1(T)}{|T|}$ for each non-empty subcoalition T of N_1. Then the largest element $T_1 \in 2^{N_1} \setminus \{\emptyset\}$ in $\arg\max_{T \in 2^{N_1} \setminus \{\emptyset\}} \frac{v_1(T)}{|T|}$ is taken (such an element exists according to Lemma 5.19) and $E_i\,(N, v) = \frac{v_1(T)}{|T|}$ for all $i \in T_1$ is defined. If $T_1 = N$, then we stop.

In case $T_1 \neq N$, then in Step 2 of the algorithm one considers the convex game $\langle N_2, v_2 \rangle$ where $N_2 := N_1 \setminus T_1$ and $v_2(S) = v_1\,(S \cup T_1) - v_1\,(T_1)$ for each $S \in 2^{N_2} \setminus \{\emptyset\}$ (cf. Lemma 5.20) takes the largest element T_2 in $\arg\max_{T \in 2^{N_2} \setminus \{\emptyset\}} \frac{v_2(T)}{|T|}$ and defines $E_i\,(v) = \frac{v_2(T)}{|T|}$ for all $i \in T_2$. If $T_1 \cup T_2 = N$ we stop; otherwise we continue by considering the game $\langle N_3, v_3 \rangle$ with $N_3 := N_2 \setminus T_2$ and $v_3(S) = v_2\,(S \cup T_2) - v_2\,(T_2)$ for each $S \in 2^{N_3} \setminus \{\emptyset\}$, etc. After a finite number of steps the algorithm stops, and the obtained allocation $E\,(v) \in \mathbb{R}^n$ is called the *constrained egalitarian solution* of the game $v \in CG^N$.

Theorem 5.22. *Let $v \in CG^N$ and let $E(v)$ be the constrained egalitarian solution. Then $E(v) \in C(v)$.*

Proof. Suppose that $S_1, \ldots, S_m$ is the ordered partition of N on which $E(v)$ is based. So,

$$E_i(v) = \frac{1}{|S_1|} v(S_1) \text{ if } i \in S_1,$$

and for $k \geq 2$:

$$E_i(v) = \frac{1}{|S_k|}\left(v\left(\cup_{r=1}^{k} S_r\right) - v\left(\cup_{r=1}^{k-1} S_r\right)\right) \text{ if } i \in S_k,$$

and for all $T \subset \cup_{r=k}^{m} S_r$ $(k \geq 1)$:

$$\frac{v\left(\left(\cup_{r=1}^{k-1} S_r\right) \cup T\right) - v\left(\cup_{r=1}^{k-1} S_r\right)}{|T|} \leq \frac{v\left(\cup_{r=1}^{k} S_r\right) - v\left(\cup_{r=1}^{k-1} S_r\right)}{|S_r|}. \quad (5.7)$$

First, we prove that $E(v)$ is efficient, i.e. $\sum_{i=1}^{n} E_i(v) = v(N)$. This follows by noting that

$$\begin{aligned}\sum_{i=1}^{n} E_i(N,v) &= \sum_{i \in S_1} E_i(v) + \sum_{k=2}^{m} \sum_{i \in S_k} E_i(v) \\ &= v(S_1) + \sum_{k=2}^{m}\left(v\left(\cup_{r=1}^{k} S_r\right) - v\left(\cup_{r=1}^{k-1} S_r\right)\right) \\ &= v\left(\cup_{r=1}^{m} S_r\right) = v(N).\end{aligned}$$

Now, we prove the stability of $E(v)$. Take $S \subset N$. We have to prove that $\sum_{i \in S} E_i(v) \geq v(S)$.

Note first that $S = \cup_{r=1}^{m} T_r$, where $T_r := S \cap S_r$ $(r = 1, \ldots, m)$. Then

$$\begin{aligned}\sum_{i \in S} E_i(v) &= \sum_{i \in T_1} E_i(v) + \sum_{k=2}^{m} \sum_{i \in T_k} E_i(v) \\ &= |T_1| \frac{v(S_1)}{|S_1|} + \sum_{k=2}^{m} |T_k| \frac{\left(v\left(\cup_{r=1}^{k} S_r\right) - v\left(\cup_{r=1}^{k-1} S_r\right)\right)}{|S_k|} \\ &\geq |T_1| \frac{v(T_1)}{|T_1|} + \sum_{k=2}^{m} |T_k| \frac{\left(v\left(\left(\cup_{r=1}^{k-1} S_r\right) \cup T_k\right) - v\left(\cup_{r=1}^{k-1} S_r\right)\right)}{|T_k|}\end{aligned}$$

$$\geq v(T_1) + \sum_{k=2}^{m} \left(v\left(\left(\cup_{r=1}^{k-1} T_r \right) \cup T_k \right) - v\left(\cup_{r=1}^{k-1} T_r \right) \right)$$
$$= v\left(\cup_{r=1}^{m} T_r \right) = v(S),$$

where the first inequality follows from (5.7), and the second inequality follows by the convexity of v by noting that $\cup_{r=1}^{k-1} S_r \supset \cup_{r=1}^{k-1} T_r$ for all $k \in \{2, \ldots, m\}$.

Since the constrained egalitarian solution is in the core of the corresponding convex game, it has been interesting to study the interrelation between $E(v)$ and every other core allocation in terms of the Lorenz criterion introduced in Section 4.1. It turns out that for convex games the constrained egalitarian solution Lorenz dominates every other core allocation; for a proof the reader is referred to [46].

Finally, we mention the population monotonicity of the constrained egalitarian solution on the domain of convex games (see [45] and [58] for details).

5.2.4 Properties of Solution Concepts

We turn now to properties of solution concepts on the class of convex games.

Notice first that it easily follows from Theorem 5.10 that for each game $v \in CG^N$ the Shapley value $\Phi(v)$ coincides with the barycenter of the core $C(v)$. Moreover, on the cone of convex games the AL-value introduced in Section 3.3 and the Shapley value coincide.

Example 5.23. Let $v \in CG^{\{1,2,3\}}$ with $v(i) = 0$ for each $i \in N$, $v(S) = 10$ if $|S| = 2$ and $v(N) = 30$. Then $L^{(1,2,3)}(v) = (20, 10, 0) = m^{(3,2,1)}(v)$, $L^{(1,3,2)}(v) = (20, 0, 10) = m^{(2,3,1)}(v), \ldots, L^{(3,2,1)}(v) = (0, 10, 20) = m^{(1,2,3)}(v)$. So, $AL(v) = (10, 10, 10) = \frac{1}{3!} \sum_{\sigma \in \pi(N)} L^{\sigma}(v) = \frac{1}{3!} \sum_{\sigma \in \pi(N)} m^{\bar{\sigma}}(v)$ $= \phi(v)$ where $\bar{\sigma} = (\sigma(3), \sigma(2), \sigma(1))$ is the reverse order of σ.

Theorem 5.24. *If $v \in CG^N$, then $AL(v) = \Phi(v)$.*

Proof. Note that for each $\sigma \in \pi(N) : L^{\sigma}(v) = m^{\bar{\sigma}}(v)$, where $\bar{\sigma} = (\sigma(n), \sigma(n-1), \ldots \sigma(2), \sigma(1))$.

Exact games (cf Definition 5.11) also play an interesting role in relation with the AL-value. Let us denote by EX^N the set of exact games with player set N. In fact, EX^N is a cone of games and one easily sees that $AL : EX^N \to \mathbb{R}^n$ is additive on subcones of EX^N, where the core correspondence is additive.

For each balanced game $v \in G^N$ there is a unique exact game $v^E \in EX^N$ with the same core as the original game. This *exactification* v^E of v is defined by $v^E(\emptyset) = 0$ and for each $S \in 2^N \setminus \{\emptyset\}$,

$$v^E(S) = \min\left\{\sum_{i \in S} x_i \mid x \in C(v)\right\}.$$

So, $C(v^E) = C(v)$ for each balanced game v and $v^E = v$ if and only if v is exact. Note the following interesting property of the AL-value: $C(v) = C(w) \neq \emptyset$ for $v, w \in G^N$ implies $AL(v) = AL(w)$. This property is equivalent to the *invariance with respect to exactification*: $AL(v) = AL(v^E)$ for each balanced game $v \in G^N$.

The invariance with respect to exactification property of the AL-value gives us the possibility to prove that for some games $v \in G^N$ the AL-value of v coincides with the Shapley value of the exactification v^E of v. This is always the case for a game $v \in G^N$ for which its exactification is convex.

Theorem 5.25. *Let $v \in G^N$.*

(i) If v is a balanced 2-person game or 3-person game, then $AL(v) = \Phi(v^E)$;
(ii) If v is a simplex game, then $AL(v) = \Phi(v^E)$;
(iii) If v is a dual simplex game, then $AL(v) = \Phi(v^E)$.

Proof. We show only (ii). Let $v \in G^N$ be a simplex game. Then $C(v) = I(v) = co\{f^1(v), \ldots, f^n(v)\}$. Further, $v^E(N) = v(N)$ and for each $S \in 2^N \setminus \{\emptyset, N\}$: $v^E(S) = \min\left\{\sum_{i \in S} x_i \mid x \in C(v)\right\} = \min\left\{\sum_{i \in S} f_i^k(v) \mid k \in \{1, \ldots, n\}\right\} = \min\left\{\sum_{i \in S} v(i), v(N) - \sum_{i \in N \setminus S} v(i)\right\} = \sum_{i \in S} v(i)$. This implies that v^E is a sum of convex games, namely $v^E = \sum_{i=1}^{n} v(i) u_{\{i\}} + \left(v(N) - \sum_{k=1}^{n} v(k)\right) u_N$ (where u_S denotes the unanimity game on S (cf. (1.1))). So, v^E is a convex game and, by Theorem 5.24, $AL(v) = AL(v^E) = \phi(v^E)$.

Now, we give a 4-person exact game v, where $\Phi(v) = \Phi(v^E) \neq AL(v)$. This game is a slight variant of an example in [39] on p. 91.

Example 5.26. Let $\varepsilon \in (0, 1]$ and let $v \in G^{\{1,2,3,4\}}$ with $v(S) = 7, 12, 22$ if $|S| = 2, 3, 4$, respectively, and $v(1) = \varepsilon$, $v(2) = v(3) = v(4) = 0$. Note that $v \notin CG^N$ because $v(1,2,3) - v(1,2) = 5 < v(1,3) - v(1) = 7 - \varepsilon$. Note further that the set of extreme points of $C(v)$ consists of (the maximum number of) 24 extreme points for a 4-person game:

(i) 12 extreme points which are permutations of $(10, 5, 5, 2)$,

(ii) 9 extreme points which are permutations of $(7, 7, 8, 0)$ but with first coordinate unequal to 0,

(iii) $(\varepsilon, 7-\varepsilon, 7-\varepsilon, 8+\varepsilon)$, $(\varepsilon, 7-\varepsilon, 8-\varepsilon, 7-\varepsilon)$ and $(\varepsilon, 8-\varepsilon, 7-\varepsilon, 7-\varepsilon)$.

From this follows that v is an exact game, and that each lexicographic maximum $L^\sigma(v)$ is equal to a permutation of the vector $(10, 5, 5, 2)$, where each such permutation corresponds to two orders. So, $AL(v) = (5\frac{1}{2}, 5\frac{1}{2}, 5\frac{1}{2}, 5\frac{1}{2})$ and is unequal to $\Phi(v^E) = \Phi(v) = (5\frac{1}{2} + \frac{1}{4}\varepsilon, 5\frac{1}{2} - \frac{1}{12}\varepsilon, 5\frac{1}{2} - \frac{1}{12}\varepsilon, 5\frac{1}{2} - \frac{1}{12}\varepsilon)$.

We show next that the equal split-off set (cf. Section 4.2) and the core (cf. Definition 2.3) of a cooperative game have additional nice properties when the game is convex. We start by proving that the equal split-off set consists of a single allocation which is the Dutta-Ray egalitarian solution of that game.

Let $\langle D_1, \ldots, D_P \rangle$ be the ordered partition of N according to the Dutta-Ray algorithm for finding the constrained egalitarian solution $E(v)$ of a game $v \in CG^N$. In each step $p \in \{1, \ldots, P\}$ of the Dutta-Ray algorithm, the coalition D_p is the largest element in the set

$$M^p := \arg \max_{S \in 2^{N \setminus \cup_{i=1}^{p-1} D_i} \setminus \{\emptyset\}} \frac{v\left(S \cup \left(\cup_{i=1}^{p-1} D_i\right)\right) - v\left(\cup_{i=1}^{p-1} D_i\right)}{|S|}.$$

We recall that for each $p \in \{1, \ldots, P\}$ the set M^p has a lattice structure w.r.t. the partial ordering of inclusion (cf. Lemma 5.19). So,

$$D_p = \cup \{D \mid D \in M^p\}.$$

For further use, we let

$$d_p := \frac{v\left(D_p \cup \left(\cup_{r=1}^{p-1} D_r\right)\right) - v\left(\cup_{r=1}^{p-1} D_r\right)}{|D_p|} \quad \text{for each } p \in \{1, \ldots, P\}.$$

Suppose now that we are given a game $v \in CG^N$, its ordered partition $\langle D_1, \ldots, D_P \rangle$ according to the Dutta-Ray algorithm, and an allocation $x = (x_i)_{i \in N}$ in the equal split-off set $ESOS(v)$ of v that is generated by the suitable ordered partition $\langle T_1, \ldots, T_K \rangle$ (cf. Definition 4.1). Then we have the following

Lemma 5.27. *Let* $\frac{v_1(T_1)}{|T_1|} = a_1$ *and* $k_1 \in \{1, \ldots, K\}$ *be the largest number for which* $\frac{v_{k_1}(T_{k_1})}{|T_{k_1}|} = a_1$. *Then* $a_1 = d_1$ *and* $D_1 = \cup_{j=1}^{k_1} T_j$.

Proof. Since $T_1 \in M^1$ and $D_1 = \cup \{D \mid D \in M^1\}$, we have $T_1 \subseteq D_1$ and $a_1 = \frac{v_1(T_1)}{|T_1|} = \frac{v(T_1)}{|T_1|} = d_1$. Next we show that $\cup_{j=1}^{k_1} T_j \subseteq D_1$ by proving by induction that $\cup_{j=1}^{k'} T_j \subseteq D_1$ for each $k' \in \{1, \ldots, k_1\}$.

For $k' = 1$ the inclusion is correct. Suppose that for some $k' \in \{1, \ldots, k_1 - 1\}$ it holds that $\cup_{j=1}^{k'} T_j \subseteq D_1$. We show that $\cup_{j=1}^{k'+1} T_j \subseteq D_1$. We have

$$\begin{aligned} d_1 \left| T_{k'+1} \right| &= v_{k'+1}(T_{k'+1}) = v_1(\cup_{j=1}^{k'+1} T_j) - v_1(\cup_{j=1}^{k'} T_j) \\ &= v_1(\cup_{j=1}^{k'+1} T_j) - d_1 \left| \cup_{j=1}^{k'} T_j \right|, \end{aligned}$$

i.e.,

$$d_1 \left| \cup_{j=1}^{k'+1} T_j \right| = v_1(\cup_{j=1}^{k'+1} T_j),$$

implying that $\cup_{j=1}^{k'+1} T_j \subseteq D_1$. Hence, $\cup_{j=1}^{k_1} T_j \subseteq D_1$.

Next we prove that $\cup_{j=1}^{k_1} T_j = D_1$. Suppose that we have $V = D_1 \setminus \cup_{j=1}^{k_1} T_j \neq \emptyset$. First, from the selection of k_1 and Proposition 4.6 it follows that

$$\frac{v_{k_1+1}(S)}{|S|} < a_1 = d_1 \quad \text{for each } S \in 2^{N \setminus \cup_{j=1}^{k_1} T_j} \setminus \{\emptyset\},$$

implying

$$\frac{v_{k_1+1}(V)}{|V|} < d_1. \tag{5.8}$$

On the other hand,

$$\begin{aligned} \frac{v_{k_1+1}(V)}{|V|} &= \frac{v_1(D_1) - v_1\left(\cup_{j=1}^{k_1} T_j\right)}{|V|} \\ &= \frac{d_1 |D_1| - d_1 \left| \left(\cup_{j=1}^{k_1} T_j\right) \right|}{|V|} = \frac{d_1 |V|}{|V|} \\ &= d_1 \end{aligned}$$

which contradicts (5.8). Hence, we have proved that $\cup_{j=1}^{k_1} T_j = D_1$.

Theorem 5.28. *Let* $v \in CG^N$. *Then* $ESOS(v) = \{E(v)\}$.

Proof. Let $\langle D_1, \ldots, D_P \rangle$ be the ordered partition of N according to the Dutta-Ray algorithm for finding the constrained egalitarian solution $E(v) = (E_i(v))_{i \in N}$ of v. Take an arbitrary allocation $x = (x_i)_{i \in N}$ in the equal split-off set $ESOS(v)$ of v, and let it be generated by the suitable ordered partition $\langle T_1, \ldots, T_K \rangle$. We show by induction on $p \in \{1, \ldots, P\}$ that there exist $k_1^*, \ldots, k_p^*, \ldots, k_P^* \in \{1, \ldots, K\}$ with $1 \leq k_1^* < \ldots < k_p^* < \ldots < k_P^* \leq K$ such that for each $p \in \{1, \ldots, P\}$ and each $j \in \left\{k_{p-1}^* + 1, \ldots, k_p^*\right\}$ we have

$$\cup_{j=k_{p-1}^*+1}^{k_p^*} T_j = D_p \text{ and } \frac{v_j(T_j)}{|T_j|} = d_p. \tag{5.9}$$

For $p = 1$, let $k_1^* = k_1$ where $k_1 \in \{1, \ldots, K\}$ is the largest number for which $\frac{v_{k_1}(T_{k_1})}{|T_{k_1}|} = a_1 = \frac{v_1(T_1)}{|T_1|}$. By Lemma 5.27 we have $\cup_{j=1}^{k_1} T_j = D_1$ and $\frac{v_j(T_j)}{|T_j|} = d_1$ for each $j \in \{1, \ldots, k_1^*\}$.

Suppose that for some $p \in \{1, \ldots, P-1\}$ there exists k_p^* such that $k_{p-1}^* < k_p^* < K$ for which (5.9) holds. We show that there exists k_{p+1}^*, such that $k_p^* < k_{p+1}^* \leq K$, for which $\cup_{j=k_p^*+1}^{k_{p+1}^*} T_j = D_{p+1}$ and $\frac{v_j(T_j)}{|T_j|} = d_{p+1}$ for each $j \in \left\{k_p^* + 1, \ldots, k_{p+1}^*\right\}$.

Notice that $\cup_{j=1}^{k_p^*} T_j = \cup_{l=1}^{p} D_l$ implying that the game $v_{k_p^*+1} : 2^{\left(N \setminus \cup_{j=1}^{k_p^*} T_j\right)} \to \mathbb{R}$ defined by

$$v_{k_p^*+1}(S) := v\left(\left(\cup_{j=1}^{k_p^*} T_j\right) \cup S\right) - v\left(\cup_{j=1}^{k_p^*} T_j\right),$$

and the game $v_{p+1} : 2^{\left(N \setminus \cup_{l=1}^{p} D_l\right)} \to \mathbb{R}$ defined by

$$v_{p+1}(S) := v\left(\left(\cup_{l=1}^{p} D_l\right) \cup S\right) - v\left(\cup_{l=1}^{p} D_l\right)$$

coincide.

Let $a_{k_p^*+1} = \frac{v_{k_p^*+1}(T_{k_p^*+1})}{\left|T_{k_p^*+1}\right|}$ and $k_{p+1} \in \left\{k_p^* + 1, \ldots, K\right\}$ be the largest number for which $\frac{v_{k_{p+1}}(T_{k_{p+1}})}{\left|T_{k_{p+1}}\right|} = a_{k_p^*+1}$. Take $k_{p+1}^* = k_{p+1}$. Given the coincidence of the games $v_{k_p^*+1}$ and v_{p+1} and their convexity (cf. [46]), we can apply the same argument as in Lemma 5.27 to conclude that (5.9) holds also for $p+1$.

It follows then that the suitable ordered partition $\langle T_1, \ldots, T_K \rangle$ is a refinement of $\langle D_1, \ldots, D_P \rangle$ of the form

$$\left\langle \langle T_1, \ldots, T_{k_1^*} \rangle, \langle T_{k_1^*+1}, \ldots, T_{k_2^*} \rangle, \ldots, \left\langle T_{k_{P-1}^*+1}, \ldots, T_{k_P^*} \right\rangle \right\rangle$$

with $T_{k_P^*} = T_K$, and for each partition $\left\langle T_{k_{p-1}^*+1}, \ldots, T_{k_p^*} \right\rangle$ of D_p, $p \in \{1, \ldots, P\}$, the members of each element T_k, $k \in \left\{k_{p-1}^* + 1, \ldots, k_p^*\right\}$ with $k_0^* = 0$, receive the same payoff d_p. Thus, we have $x = E(v)$ implying that $ESOS(v) = \{E(v)\}$ for $v \in CG^N$.

Finally, with respect to convex games the following result is well known.

Theorem 5.29. *([104]) Let $v \in CG^N$. Then $C(v)$ is the unique stable set.*

Proof. In view of Theorem 2.12 we only have to show that $C(v)$ is stable. This is true if v is additive. So we suppose that v is not additive.

Let $y \in I(v) \setminus C(v)$. Take an $S \in 2^N \setminus \{\emptyset\}$ such that

$$\frac{v(S) - \sum_{i \in S} y_i}{|S|} = \max_{C \in 2^N \setminus \{\emptyset\}} \frac{v(C) - \sum_{i \in C} y_i}{|C|}. \tag{5.10}$$

Further take $z \in C(v)$ such that $\sum_{i \in S} z_i = v(S)$. This is possible in view of Remark 5.12.

Let $x \in \mathbb{R}^n$ be the vector with

$$x_i := \begin{cases} y_i + \frac{v(S) - \sum_{i \in S} y_i}{|S|} & \text{if } i \in S, \\ z_i & \text{otherwise.} \end{cases}$$

Then $x \in I(v)$ and $x \operatorname{dom}_S y$. To prove that $x \in C(v)$, note first of all that for $T \in 2^N$ with $T \cap S \neq \emptyset$ we have

$$\begin{aligned} \sum_{i \in T \cap S} x_i &= \sum_{i \in T \cap S} (x_i - y_i) + \sum_{i \in T \cap S} y_i \\ &= |T \cap S| \frac{v(S) - \sum_{i \in S} y_i}{|S|} + \sum_{i \in T \cap S} y_i \\ &\geq \left(v(T \cap S) - \sum_{i \in T \cap S} y_i \right) + \sum_{i \in T \cap S} y_i \\ &= v(T \cap S), \end{aligned}$$

where the inequality follows from (5.10).

But then

$$\begin{aligned}\sum_{i\in T} x_i &= \sum_{i\in T\cap S} x_i + \sum_{i\in T\setminus S} z_i \\ &\geq v(T\cap S) + \sum_{i\in T\cup S} z_i - \sum_{i\in S} z_i \\ &\geq v(T\cap S) + v(T\cup S) - v(S) \\ &\geq v(T)\end{aligned}$$

because $z \in C(v)$, $\sum_{i\in S} z_i = v(S)$ and $v \in CG^N$.

For $T \in 2^N \setminus \{\emptyset\}$ with $T \cap S = \emptyset$ we have $\sum_{i\in T} x_i = \sum_{i\in T} z_i \geq v(T)$ because $z \in C(v)$. So we have proved that $x \in C(v)$.

Then $I(v) = C(v) \cup dom\,(C(v))$ and $C(v) \cap dom(C(v)) = \emptyset$. Hence, $C(v)$ is a stable set.

5.3 Clan Games

Clan games were introduced in [89] to model social conflicts between "powerful" players (clan members) and "powerless" players (non-clan members). In a clan game the powerful players have veto power and the powerless players operate more profitably in unions than on their own. Economic applications of such clan games to bankruptcy problems, production economies, holding situations and information acquisition are provided in [27], [77], [89], [117], and [122].

5.3.1 Basic Characterizations and Properties of Solution Concepts

Definition 5.30. *A game $v \in G^N$ is a* **clan game** *with clan $C \in 2^N \setminus \{\emptyset, N\}$ if it satisfies the following four conditions:*

(a) Nonnegativity: $v(S) \geq 0$ for all $S \subset N$;
(b) Nonnegative marginal contributions to the grand coalition: $M_i(N,v) \geq 0$ for each player $i \in N$;
(c) Clan property: every player $i \in C$ is a veto player, i.e. $v(S) = 0$ for each coalition S that does not contain C;
(d) Union property: $v(N) - v(S) \geq \sum_{i\in N\setminus S} M_i(N,v)$ if $C \subset S$.

If the clan consists of a single member, the corresponding game is called a *big boss game* (cf. [77]).

The next proposition shows that the core of a clan game has an interesting shape.

Proposition 5.31. *([89]) Let $v \in G^N$ be a clan game. Then*

$$C(v) = \{x \in I(v) \mid x_i \leq M_i(N,v) \text{ for all } i \in N \setminus C\}.$$

Proof. Suppose $x \in C(v)$. Then $\sum_{j \in N \setminus \{i\}} x_j \geq v(N \setminus \{i\})$ for all $i \in N \setminus C$. Since $v(N) = \sum_{i \in N} x_i = \sum_{j \in N \setminus \{i\}} x_j + x_i$ one has

$$\begin{aligned} x_i = v(N) - \sum_{j \in N \setminus \{i\}} x_j &\leq v(N) - v(N \setminus \{i\}) \\ &= M_i(N,v) \text{ for all } i \in N \setminus C. \end{aligned}$$

Conversely, if $x \in I(v)$ and $x_i \leq M_i(N,v)$ for all $i \in N \setminus C$, then, for a coalition S which does not contain C, one finds that $\sum_{i \in S} x_i \geq 0 = v(S)$. If $C \subset S$, then, by using condition (d) in Definition 5.30, one has

$$v(N) - v(S) \geq \sum_{i \in N \setminus S} M_i(N,v) \geq \sum_{i \in N \setminus S} x_i.$$

Since $v(N) = \sum_{i \in N} x_i$, one finally obtains $\sum_{i \in S} x_i \geq v(S)$, i.e. $x \in C(v)$.

In fact, a clan game can be fully described by the shape of the core as indicated in

Proposition 5.32. *([89]) Let $v \in G^N$ and $v \geq 0$. The game v is a clan game iff*
(i) $v(N)e^j \in C(v)$ for all $j \in C$;
(ii) There is at least one element $x \in C(v)$ such that $x_i = M_i(N,v)$ for all $i \in N \setminus C$.

Proof. One needs to prove only sufficiency. Suppose $S \in 2^N \setminus \{\emptyset\}$ does not contain C. Take $j \in C \setminus S$. Because $x := v(N)e^j \in C(v)$, one has $\sum_{i \in S} x_i = 0 \geq v(S)$ and from $v \geq 0$ one finds the clan property in Definition 5.30.

If $C \subset S$ and $x \in C(v)$ with $x_i = M_i(N,v)$ for all $i \in N \setminus C$, then

$$v(S) \leq \sum_{i \in S} x_i = \sum_{i \in N} x_i - \sum_{i \in N \setminus S} x_i = v(N) - \sum_{i \in N \setminus S} M_i(N,v),$$

proving the union property in Definition 5.30.

Furthermore, $M_i(N,v) = x_i \geq v(i) = 0$ for all $i \in N \setminus C$.

Now, we focus on the AL-value (cf. Section 3.3) for big boss games. Note that, according to Proposition 5.31, the extreme points of the core of a big boss game v with n as big boss are of the form P^T where $T \subseteq N \setminus \{n\}$ and $P_i^T = M_i(v)$ if $i \in T$, $P_i^T = 0$ if $i \in N \setminus (T \cup \{n\})$ and $P_n^T = v(N) - \sum_{i \in T} M_i(v)$. For each $\sigma \in \pi(N)$ the lexicographic maximum $L^\sigma(v)$ equals $P^{T(\sigma)}$, where $T(\sigma) = \{i \in N \setminus \{n\} \mid \sigma(i) < \sigma(n)\}$.

Theorem 5.33. *Let $v \in G^N$ be a big boss game with n as big boss. Then $AL(v) = \tau(v)$.*

Proof. For each $i \in N \setminus \{n\}$ we have $AL_i(v) = \frac{1}{n!} \sum_{\sigma \in \pi(N)} (L^\sigma(v))_i = \frac{1}{n!} \sum_{\sigma \in \pi(N)} (P^{T(\sigma)})_i = \frac{1}{n!} M_i(v) |\{\sigma \in \pi(N) \mid \sigma(i) < \sigma(n)\}| = \frac{1}{2} M_i(v) = \tau_i(v)$.

Finally, since the τ-value and the AL-value are efficient, we obtain that $AL_n(v) = \tau_n(v)$, too.

Let us now look at the exactification v^E (cf. Section 5.2.1) of the big boss game v with n as a big boss. For $S \subseteq N \setminus \{n\}$ we have

$$v^E(S) = \min_{T \subseteq N \setminus \{n\}} \sum_{i \in S} P_i^T = \sum_{i \in S} P_i^\Phi = 0,$$

while for $S \subseteq N$ with $n \in S$ we have

$$\begin{aligned} v^E(S) &= \min\nolimits_{T \subseteq N \setminus \{n\}} \sum_{i \in S} P_i^T = \min\nolimits_{T \subseteq N \setminus \{n\}} (v(N) - \sum_{i \in T \setminus S} M_i(v)) \\ &= \sum_{i \in S} P_i^{N \setminus \{n\}} = (v(N) - \sum_{i=1}^{n-1} M_i(v)) + \sum_{i \in S} M_i(v). \end{aligned}$$

This implies that v^E is a non-negative combination of convex unanimity games:

$$v^E = \left(v(N) - \sum_{i=1}^{n-1} M_i(v) \right) u_{\{n\}} + \sum_{i \in N \setminus \{n\}} M_i(v) u_{\{i,n\}}.$$

So, v^E is a convex game (and also a big boss game) and the extreme points of $C(v)$ and of $C(v^E)$ coincide. We obtain then that $\tau(v) = AL(v) = AL(v^E) = \Phi(v^E)$.

Theorem 5.34. *Let $v \in G^N$ be a big boss game with n as big boss. Then $AL(v) = \Phi(v^E)$.*

5.3.2 Total Clan Games and Monotonic Allocation Schemes

The subgames in a total clan game inherit the structure of the original (clan) game. This leads to the following

Definition 5.35. *A game $v \in G^N$ is a* **total clan game** *with clan $C \in 2^N \setminus \{\emptyset, N\}$ if v_S is a clan game (with clan C) for every coalition $S \supset C$.*

Note that in Definition 5.35 attention is restricted to coalitions that contain the clan C, since the clan property of v implies that in the other subgames the characteristic function is simply the zero function.

The next theorem provides a characterization of total clan games. The reader is referred to [122] for its proof.

Theorem 5.36. *Let $v \in G^N$ and $C \in 2^N \setminus \{\emptyset, N\}$. The following claims are equivalent:*

(i) v is a total clan game with clan C;
(ii) v is monotonic, every player $i \in C$ is a veto player, and for all coalitions S and T with $S \supset C$ and $T \supset C$:

$$S \subset T \text{ implies } v(T) - v(S) \geq \sum_{i \in T \setminus S} M_i(T, v); \tag{5.11}$$

(iii) v is monotonic, every player $i \in C$ is a veto player, and for all coalitions S and T with $S \supset C$ and $T \supset C$:

$$S \subset T \text{ and } i \in S \setminus C \text{ imply } M_i(S, v) \geq M_i(T, v). \tag{5.12}$$

One can study also the question whether a total clan game possesses a pmas (cf. Definitions 5.5, 5.6). The attention in [122] is restricted to the allocation of $v(S)$ for coalitions $S \supset C$, since other coalitions have value zero by the clan property.

Theorem 5.37. *([122]) Let $v \in G^N$ be a total clan game with clan $C \in 2^N \setminus \{\emptyset, N\}$ and let $b \in C(v)$. Then b is pmas extendable.*

Proof. According to Proposition 5.31 we have

$$C(v) = \{x \in I(v) \mid x_i \leq M_i(N, v) \text{ for all } i \in N \setminus C\}.$$

Hence, there exists, for each player $i \in N$, a number $\alpha_i \in [0, 1]$ such that $\sum_{i \in C} \alpha_i = 1$ and

$$b_i = \begin{cases} \alpha_i M_i(N, v) & \text{if } i \in N \setminus C, \\ \alpha_i \left[v(N) - \sum_{j \in N \setminus C} \alpha_j M_j(N, v) \right] & \text{if } i \in C. \end{cases}$$

In other words, each non-clan member receives a fraction of his marginal contribution to the grand coalition, whereas the clan members divide the remainder.

Define for each $S \supset C$ and $i \in S$:

$$a_{iS} = \begin{cases} \alpha_i M_i(N,v) & \text{if } i \in S \setminus C, \\ \alpha_i \left[v(S) - \sum_{j \in S \setminus C} \alpha_j M_j(N,v)\right] & \text{if } i \in C. \end{cases}$$

Clearly, $a_{iN} = b_i$ for each player $i \in N$. We proceed to prove that the vector $(a_{iS})_{i \in S, S \supset C}$ is a pmas. Since $\sum_{i \in C} \alpha_i = 1$, it follows that $\sum_{i \in S} a_{iS} = v(S)$. Now let $S \supset C, T \supset C$ and $i \in S \subset T$.
- If $i \notin C$, then $a_{iS} = a_{iT} = \alpha_i M_i(N,v)$.
- If $i \in C$, then

$$\begin{aligned} &a_{iT} - a_{iS} \\ &= \alpha_i \left[v(T) - \sum_{j \in T \setminus C} \alpha_j M_j(N,v)\right] - \alpha_i \left[v(S) - \sum_{j \in S \setminus C} \alpha_j M_j(N,v)\right] \\ &= \alpha_i \left[v(T) - v(S) - \sum_{j \in T \setminus S} \alpha_j M_j(N,v)\right] \\ &\geq \alpha_i \left[v(T) - v(S) - \sum_{j \in T \setminus S} M_j(N,v)\right] \\ &\geq \alpha_i \left[v(T) - v(S) - \sum_{j \in T \setminus S} M_j(T,v)\right] \\ &\geq 0, \end{aligned}$$

where the first inequality follows from nonnegativity of the marginal contributions and the fact that $\alpha_j \leq 1$ for each $j \in N \setminus C$, the second inequality follows from (5.12), and the final inequality from (5.11) and nonnegativity of α_i for each $i \in C$. Consequently, $(a_{iS})_{i \in S, S \supset C}$ is a pmas.

Whereas a pmas allocates a larger payoff to each player as the coalitions grow larger, property (5.12) suggests a slightly different approach in total clan games: the marginal contribution of each non-clan member actually decreases in a larger coalition. Taking this into account, one might actually allocate a smaller amount to the non-clan members in larger coalitions. Moreover, to still maintain some stability, such allocations should still give rise to core allocations in the subgames. An allocation scheme that satisfies these properties is called *bi-monotonic allocation scheme* (cf. [122]).

Definition 5.38. *Let $v \in G^N$ be a total clan game with clan $C \in 2^N \setminus \{\emptyset, N\}$. A* **bi-monotonic allocation scheme** *(bi-mas) for the game v is a vector $a = (a_{iS})_{i\in S, S\supset C}$ of real numbers such that*

(i) $\sum_{i\in S} a_{iS} = v(S)$ *for all* $S \in 2^C \setminus \{\emptyset\}$,

(ii) $a_{iS} \leq a_{iT}$ *for all* $S \supset C, T \supset C$ *with* $S \subset T$ *and* $i \in S \cap C$,

(iii) $a_{iS} \geq a_{iT}$ *for all* $S \supset C, T \supset C$ *with* $S \subset T$ *and* $i \in S \setminus C$,

(iv) $(a_{iS})_{i\in S}$ *is a core element of the subgame* v_S *for each coalition* $S \supset C$.

Definition 5.39. *Let $v \in G^N$ be a total clan game with clan $C \in 2^N \setminus \{\emptyset, N\}$. An imputation $b \in I(v)$ is* **bi-mas extendable** *if there exist a bi-mas $a = (a_{iS})_{i\in S, S\supset C}$ such that $a_{iN} = b_i$ for each player $i \in N$.*

Theorem 5.40. *([122]) Let $v \in G^N$ be a total clan game with clan $C \in 2^N \setminus \{\emptyset, N\}$ and let $b \in C(v)$. Then b is bi-mas extendable.*

Proof. Take $(\alpha_i)_{i\in N} \in [0,1]^N$ as in the proof of Theorem 5.37. Define for each $S \supset C$ and $i \in S$:

$$a_{iS} = \begin{cases} \alpha_i M_i(S,v) & \text{if } i \in S \setminus C, \\ \alpha_i \left[v(S) - \sum_{j\in S\setminus C} \alpha_j M_j(S,v) \right] & \text{if } i \in C. \end{cases}$$

We proceed to prove that $(a_{iS})_{i\in S, S\supset C}$ is a bi-mas. Since $\sum_{i\in C} \alpha_i = 1$, it follows that $\sum_{i\in S} a_{iS} = v(S)$. Now let $S \supset C, T \supset C$ and $i \in S \subset T$.
- If $i \in N \setminus C$, then $a_{iS} = \alpha_i M_i(S,v) \geq \alpha_i M_i(T,v) = a_{iT}$ by (5.12).
- If $i \in C$, then

$$\begin{aligned} &a_{iT} - a_{iS} \\ &= \alpha_i \left[v(T) - \sum_{j\in T\setminus C} \alpha_j M_j(T,v) \right] - \alpha_i \left[v(S) - \sum_{j\in S\setminus C} \alpha_j M_j(S,v) \right] \end{aligned}$$

$$
\begin{aligned}
&= \alpha_i \left[v(T) - v(S) - \sum_{j \in T \setminus S} \alpha_j M_j(T,v) \right] \\
&\quad + \alpha_i \left[\sum_{j \in S \setminus C} \alpha_j \left(M_j(S,v) - M_j(T,v) \right) \right] \\
&\geq \alpha_i \left[v(T) - v(S) - \sum_{j \in T \setminus S} \alpha_j M_j(T,v) \right] \\
&\geq 0
\end{aligned}
$$

where the first inequality follows from (5.12) and nonnegativity of $(\alpha_j)_{j \in T \setminus S}$, and the second inequality follows from (5.11) and nonnegativity of α_i for each $i \in C$.

Finally, for each coalition $S \supset C$, the vector $(a_{iS})_{i \in S, S \supset C}$ is shown to be a core allocation of the clan game v_S. Let $S \supset C$. According to Proposition 5.31 we have

$$C(v) = \{x \in I(v) \mid x_i \leq M_i(N,v) \text{ for all } i \in N \setminus C\}.$$

Let $i \in S \setminus C$. Then $a_{iS} = \alpha_i M_i(S,v) \leq M_i(S,v)$. Also, $\sum_{i \in S} a_{iS} = v(S)$, so $(a_{iS})_{i \in S}$ satisfies efficiency.

To prove individual rationality, consider the following three cases:

- Let $i \in S \setminus C$. Then $a_{iS} = \alpha_i M_i(S,v) \geq 0 = v(i)$;
- Let $i \in S \cap C$ and $|C| = 1$. Then $C = \{i\}$ and by construction $\alpha_i = \sum_{j \in C} \alpha_j = 1$. Hence $a_{iS} \geq a_{iC} = \alpha_i v(C) = v(i)$;
- Let $i \in S \cap C$ and $|C| > 1$. Then $a_{iS} \geq a_{iC} = \alpha_i v(C) \geq 0 = v(i)$, since every player in C is a veto player.

Consequently, $(a_{iS})_{i \in S, S \supset C}$ is a bi-mas.

5.4 Convex Games versus Clan Games

In this section we identify common features of the class of convex games and the class of clan games ([22]). We start by showing that each game in the corresponding class can be characterized by means of certain properties of appropriately defined marginal games. Further, we study the duality between general convex games and total clan games and show that, starting with a total clan game with zero worth for the clan we make use of its dual game and restrict it to the non-clan members in order to induce a monotonic convex game. Conversely, we can start with a monotonic convex game, use its dual game and assign zero worth to each coalition not containing a certain group of players as to reach a total clan game with zero worth for the clan. Finally, the way in which the corresponding games are constructed ("dualize and restrict" versus "dualize and extend") is also useful for providing relations between core elements and elements of the Weber set of the corresponding games.

Since the player set on which a game v is played will be of special importance for what follows, we write (N, v) for $v \in G^N$.

5.4.1 Characterizations via Marginal Games

Given a game (N, v) and a coalition $T \subseteq N$, the *T-marginal game* $v^T : 2^{N \setminus T} \to \mathbb{R}$ is defined by

$$v^T(S) := v(S \cup T) - v(T)$$

for each $S \subseteq N \setminus T$ (see also (4.1)).

As we already know if a game is convex then all its marginal games are also convex (cf. Lemma 5.20). The next example shows that the superadditivity (cf. Definition 1.14) of a game is not necessarily inherited by its marginal games.

Example 5.41. Let $N = \{1, 2, 3\}$ and $v(\{1\}) = 10$, $v(\{1, 2\}) = 12$, $v(\{1, 3\}) = 11$, $v(\{1, 2, 3\}) = 12\frac{1}{2}$, and $v(S) = 0$ for all other $S \subset N$. Clearly, the game (N, v) is superadditive. Its $\{1\}$-marginal game is given by $v^{\{1\}}(\{2\}) = v(\{1, 2\}) - v(\{1\}) = 2$, $v^{\{1\}}(\{3\}) = 11 - 10 = 1$, and $v^{\{1\}}(\{2, 3\}) = 2\frac{1}{2}$. Since $v^{\{1\}}(\{2, 3\}) = 2\frac{1}{2} < 3 = v^{\{1\}}(\{2\}) + v^{\{1\}}(\{3\})$, the marginal game $\left(\{2, 3\}, v^{\{1\}}\right)$ is not superadditive.

As it turns out, the superadditivity of all marginal games of a game (N, v) assures a stronger property than the superadditivity of (N, v), namely the convexity of (N, v). This result, which we give without proof, has been independently obtained in [20] and [71].

Proposition 5.42. *A game (N, v) is convex if and only if for each $T \in 2^N$ the T-marginal game $\left(N \setminus T, v^T\right)$ is superadditive.*

Remark 5.43. In view of Proposition 5.42, a game (N, v) is concave (cf. Definition 5.9) if and only if for each $T \in 2^N$ the marginal game $\left(N \setminus T, v^T\right)$ is subadditive (cf. Definition 1.15).

Let us now focus on total clan games (cf. Definition 5.35) and their characterization using suitably defined marginal games. For a clan game (N, v) with clan $C \in 2^N \setminus \{\emptyset, N\}$, define $\mathcal{P}^C := \{S \subseteq N \mid C \subseteq S\}$ as the collection of all coalitions containing C. By Theorem 5.36(iii), a game (N, v) is a total clan game with clan C if and only if (N, v) is monotonic, every player $i \in C$ is a veto player, and for all coalitions $S_1, S_2 \in \mathcal{P}^C$ the following *C-concavity property* holds:

$$S_1 \subseteq S_2 \text{ and } i \in S_1 \setminus C \text{ imply } M_i(S_1, v) \geq M_i(S_2, v).$$

In what follows we denote by $MV^{N,C}$ the set of all monotonic games on N satisfying the veto player property with respect to each player $i \in C$.

Proposition 5.44. *Let $(N, v) \in MV^{N,C}$. Then (N, v) is a total clan game with clan C if and only if the marginal game $\left(N \setminus C, v^C\right)$ is a concave game.*

Proof. Let $(N, v) \in MV^{N,C}$ be a total clan game with clan C. Then for $i \in S \subseteq T \subseteq N$ we have

$$\begin{aligned} v^C(S) - v^C(S \setminus \{i\}) &= v(C \cup S) - v((C \cup S) \setminus \{i\}) \\ &\geq v(C \cup T) - v((C \cup T) \setminus \{i\}) \\ &= v^C(T) - v^C(T \setminus \{i\}), \end{aligned}$$

where the inequality follows from the C-concavity of (N, v). Hence, (N, v) is a concave game.

Suppose now that $\left(N \setminus C, v^C\right)$ is a concave game. Let $S_1, S_2 \in \mathcal{P}^C$, $S_1 \subseteq S_2$, and $i \in S_1 \setminus C$. Then

$$\begin{aligned} M_i(S_1, v) &= v(S_1) - v(S_1 \setminus \{i\}) \\ &= v^C(S_1 \setminus C) - v^C((S_1 \setminus C) \setminus \{i\}) \\ &\geq v^C(S_2 \setminus C) - v^C((S_2 \setminus C) \setminus \{i\}) \\ &= M_i(S_2, v), \end{aligned}$$

where the inequality follows from the concavity of $\left(N \setminus C, v^C\right)$. Thus, (N, v) is a total clan game with clan C. $\square$

Given a game $(N, v) \in MV^{N,C}$ and a coalition $T \in 2^{N \setminus C}$, the *C-based T-marginal game* $\left(v^C\right)^T : 2^{N \setminus T} \to \mathbb{R}$ is defined by

$$\left(v^C\right)^T(S) := v(S \cup T \cup C) - v(T \cup C)$$

for each $S \subseteq N \setminus T$.

We have then the following result.

Proposition 5.45. *Let $(N, v) \in MV^{N,C}$. Then the following assertions are equivalent:*

(a) (N, v) *is a total clan game with clan* C*;*
(b) $\left(N \setminus C, v^C\right)$ *is a concave game;*
(c) $\left(N \setminus (C \cup T), \left(v^C\right)^T\right)$ *is a subadditive game for each* $T \subseteq N \setminus C$*;*
(d) $\left(N \setminus (C \cup T), v^{C \cup T}\right)$ *is a subadditive game for each* $T \subseteq N \setminus C$*.*

Proof. Notice that (a) $\Longleftrightarrow$ (b) follows from Proposition 5.44, and (b) $\Longleftrightarrow$ (c) holds by Remark 5.43. Finally, (c) $\Longleftrightarrow$ (d) follows easily from the definition of a C-based T-marginal game.

5.4.2 Dual Transformations

We present now a useful relation between total clan games with zero worth for the clan and monotonic convex games, being interested in transformations that work across these two classes of games. As it turns out, we can always construct monotonic convex games from total clan games with zero worth for the clan, and total clan games with zero worth for the clan from monotonic convex games. We call the corresponding transformation procedures "dualize and restrict" and "dualize and extend", respectively.

Let $N = \{1, \ldots, n\}$ and $C \in 2^N \setminus \{\emptyset, N\}$ be fixed. We denote by $CLAN_0^{N,C}$ the set of all total clan games on N with clan C for which $v(C) = 0$ is valid. The set of all games on $N \setminus C$ will be denoted by $G^{N \setminus C}$ and the set of all monotonic convex games on $N \setminus C$ by $MCONV^{N \setminus C}$.

The *"dualize and restrict" operator* $D^r : CLAN_0^{N,C} \to G^{N \setminus C}$ is defined by

$$D^r(N, v) = (N \setminus C, w) \ \text{ for each } \ (N, v) \in CLAN_0^{N,C},$$

where $w(S) := v^*(S)$ for all $S \subseteq N \setminus C$ (cf. Definition 1.10).

Proposition 5.46. *Let $(N, v) \in CLAN_0^{N,C}$. Then*

$$D^r(N, v) \in MCONV^{N \setminus C}.$$

Proof. Let $(N \setminus C, w) := D^r(N, v)$. To show that $(N \setminus C, w)$ is convex, let $i \in N \setminus C$ and $S_1 \subseteq S_2 \subseteq (N \setminus C) \setminus \{i\}$. Then

$$\begin{aligned} w(S_2 \cup \{i\}) - w(S_2) &= v(N) - v(N \setminus (S_2 \cup \{i\})) - v(N) + v(N \setminus S_2) \\ &= v(N \setminus S_2) - v(N \setminus (S_2 \cup \{i\})) \\ &\geq v(N \setminus S_1) - v(N \setminus (S_1 \cup \{i\})) \\ &= v(N) - v(N \setminus (S_1 \cup \{i\})) - v(N) + v(N \setminus S_1) \\ &= w(S_1 \cup \{i\}) - w(S_1), \end{aligned}$$

where the inequality follows from the C-concavity of (N, v).

By the monotonicity property of (N, v) we have that $w(S_1) = v(N) - v(N \setminus S_1) \leq v(N) - v(N \setminus S_2) = w(S_2)$ for $S_1 \subseteq S_2 \subseteq N \setminus C$, i.e., the game $(N \setminus C, w)$ is monotonic as well.

The *"dualize and extend" operator* $D^e : MCONV^{N\setminus C} \to G^N$ is defined by

$$D^e(N \setminus C, w) = (N, v) \text{ for each } (N \setminus C, w) \in MCONV^{N\setminus C},$$

where

$$v(S) = \begin{cases} 0 & \text{if } C \not\subseteq S, \\ w(N \setminus C) - w((N \setminus C) \setminus (S \cap (N \setminus C))) & \text{otherwise,} \end{cases}$$

for all $S \subseteq N$.

Proposition 5.47. *Let* $(N \setminus C, w) \in MCONV^{N\setminus C}$. *Then*

$$D^e(N \setminus C, w) \in CLAN_0^{N,C}.$$

Proof. Notice that, by the definition of v, each player $i \in C$ is a veto player in the game $(N, v) := D^e(N \setminus C, w)$. To prove that (N, v) is monotonic, let $S \subset N$ and $i \in N \setminus S$.

If $C \not\subseteq S$ and either $i \notin C$, or $i \in C$ is such that $C \not\subseteq S \cup \{i\}$, then $v(S) = 0 = v(S \cup \{i\})$ follows by the definition of v.

If $C \not\subseteq S$ and $i \in C$ is such that $C \subseteq S \cup \{i\}$, then we have

$$\begin{aligned} v(S \cup \{i\}) &= w(N \setminus C) - w((N \setminus C) \setminus ((S \cup \{i\}) \cap (N \setminus C))) \\ &= w(N \setminus C) - w((N \setminus C) \setminus (S \cap (N \setminus C))) \\ &= v(S). \end{aligned}$$

If $C \subseteq S$, then

$$\begin{aligned} v(S \cup \{i\}) &= w(N \setminus C) - w((N \setminus C) \setminus ((S \cup \{i\}) \cap (N \setminus C))) \\ &\geq w(N \setminus C) - w((N \setminus C) \setminus (S \cap (N \setminus C))) \\ &= v(S), \end{aligned}$$

where the inequality follows by the monotonicity of $(N \setminus C, w)$.

It remains to show that (N, v) is C-concave. For this, let $S_1 \subseteq S_2 \subseteq N$ and $i \in S_1 \setminus C$. Then

$$\begin{aligned}
&v(S_1) - v(S_1 \setminus \{i\}) \\
&= w((N \setminus C) \setminus ((S_1 \setminus \{i\}) \cap (N \setminus C))) - w((N \setminus C) \setminus (S_1 \cap (N \setminus C))) \\
&\geq w((N \setminus C) \setminus ((S_2 \setminus \{i\}) \cap (N \setminus C))) - w((N \setminus C) \setminus (S_2 \cap (N \setminus C)))) \\
&= v(S_2) - v(S_2 \setminus \{i\}),
\end{aligned}$$

where the inequality follows by the convexity of $(N \setminus C, w)$.

Notice finally that it is straightforward to prove the following result.

Proposition 5.48. *Let D^r and D^e be the "dualize and restrict" and the "dualize and extend" operators, respectively, as introduced above. Then,*

(a) $D^e \circ D^r$ is the identity map on $CLAN_0^{N,C}$, and
(b) $D^r \circ D^e$ is the identity map on $MCONV^{N \setminus C}$.

5.4.3 The Core versus the Weber Set

We now show how to use the "dualize and restrict" and the "dualize and extend" procedures to relate core elements and elements of the Weber set (cf. Definitions 2.3 and 2.19) of corresponding (total clan and monotonic convex) games.

In order to state our results, we will need some additional notation. Let the player set N and $C \in 2^N \setminus \{\emptyset, N\}$ be fixed, and let $\Pi(C)$ and $\Pi(N \setminus C)$ denote the set of all permutations of C and $N \setminus C$, respectively. For each $(\tau, \sigma) \in \Pi(C) \times \Pi(N \setminus C)$, we write $m^{(\tau,\sigma)}(N, v)$ to denote the marginal contribution vector with respect to $(N, v) \in CLAN_0^{N,C}$ and to the permutation (τ, σ) of N according to which the set of all predecessors of each non-clan member includes the clan. We let

$$W'(N, v) := co\left\{m^{(\tau,\sigma)}(N, v) \mid (\tau, \sigma) \in \Pi(C) \times \Pi(N \setminus C)\right\}.$$

Finally, let $m^{(\tau,\sigma)}(N, v)_{|N \setminus C}$ denote the projection of $m^{(\tau,\sigma)}(N, v)$ on $N \setminus C$ and

$$W'(N, v)_{|N \setminus C} := co\left\{m^{(\tau,\sigma)}(N, v)_{|N \setminus C} \mid (\tau, \sigma) \in \Pi(C) \times \Pi(N \setminus C)\right\}.$$

We have the following result.

Proposition 5.49. *Let* $(N,v) \in CLAN_0^{N,C}$. *Then* $Core\,(D^r\,(N,v)) = W'\,(N,v)_{|N\setminus C}$.

Proof. Let $(N \setminus C, w) := D^r\,(N,v)$. By Proposition 5.46, $(N \setminus C, w) \in MCONV^{N\setminus C}$ and, hence, $Core\,(N \setminus C, w) = W\,(N \setminus C, w)$. Thus, it is sufficient to prove that $W\,(N \setminus C, w) = W'\,(N,v)_{|N\setminus C}$. For this, we show that $m^{\sigma}\,(N \setminus C, w) = m^{(\tau,\overline{\sigma})}\,(N,v)_{|N\setminus C}$ for each $\sigma \in \Pi\,(N \setminus C)$ and any $\tau \in \Pi(C)$, where

$$m^{\sigma}\,(N \setminus C, w) \in W\,(N \setminus C, w)\,,$$

$$m^{(\tau,\overline{\sigma})}\,(N,v)_{|N\setminus C} \in W'\,(N,v)_{|N\setminus C}\,,$$

and $\overline{\sigma}$ is the reverse order of σ.

For each $i \in N \setminus C$ we have

$$\begin{aligned}
m_i^{(\tau,\overline{\sigma})}\,(N,v)_{|N\setminus C} &= v\left(P^{(\tau,\overline{\sigma})}\,(i) \cup \{i\}\right) - v\left(P^{(\tau,\overline{\sigma})}\,(i)\right) \\
&= v\left(C \cup P^{\overline{\sigma}}\,(i) \cup \{i\}\right) - v\left(C \cup P^{\overline{\sigma}}\,(i)\right) \\
&= v\left(N \setminus P^{\sigma}\,(i)\right) - v\left(N \setminus (P^{\sigma}\,(i) \cup \{i\})\right) \\
&= v(N) - v\left(N \setminus (P^{\sigma}\,(i) \cup \{i\})\right) - v(N) \\
&\quad + v\left(N \setminus P^{\sigma}\,(i)\right) \\
&= w\left(P^{\sigma}\,(i) \cup \{i\}\right) - w\left(P^{\sigma}\,(i)\right) \\
&= m_i^{\sigma}\,(N \setminus C, w)\,.
\end{aligned}$$

Thus, the assertion follows.

Let $m^{\sigma}\,(N \setminus C, w)$ be the marginal contributions vector with respect to $(N \setminus C, w)$ and $\sigma \in \Pi\,(N \setminus C)$. In what follows, let $x^{m^{\sigma}(N\setminus C,w)} \in \mathbb{R}^n$ be defined by

$$x_i^{m^{\sigma}(N\setminus C,w)} = \begin{cases} 0 & \text{if } i \in C, \\ m_i^{\sigma}\,(N \setminus C, w) & \text{if } i \in N \setminus C, \end{cases}$$

and let

$$X^{(N\setminus C,w)} = co\left\{x^{m^{\sigma}(N\setminus C,w)} \mid \sigma \in \Pi\,(N \setminus C)\right\}.$$

We are ready now to present our last result.

Proposition 5.50. *Let* $(N \setminus C, w) \in MCONV^{N\setminus C}$. *Then* $W'\,(D^e\,(N \setminus C, w)) = X^{(N\setminus C,w)}$.

Proof. Let $(N, v) := D^e (N \setminus C, w)$. Since, by Proposition 5.47, $(N, v) \in CLAN_0^{N,C}$, it is sufficient to show that $x^{m^\sigma(N\setminus C,w)} = m^{(\tau,\overline{\sigma})} (N, v)$ for each $\sigma \in \Pi (N \setminus C)$ and any $\tau \in \Pi(C)$, where $\overline{\sigma}$ is the reverse order of σ. We distinguish two cases:

(a) $i \in N \setminus C$. In view of Proposition 5.49,

$$m_i^{(\tau,\overline{\sigma})} (N, v) = m_i^{(\tau,\overline{\sigma})} (N, v)_{|N\setminus C} = m_i^\sigma (N \setminus C, w) = x_i^{m^\sigma(N\setminus C,w)}.$$

(b) $i \in C$. We have

$$m_i^{(\tau,\overline{\sigma})} (N, v) = v \left(P^{(\tau,\overline{\sigma})} (i) \cup \{i\}\right) - v \left(P^{(\tau,\overline{\sigma})} (i)\right) = 0 = x_i^{m^\sigma(N\setminus C,w)},$$

where the second equality follows from $(N, v) \in CLAN_0^{N,C}$ and $P^{(\tau,\overline{\sigma})} (i) \cup \{i\} \subseteq C$.

Hence, the assertion follows.

Part II

Cooperative Games with Fuzzy Coalitions

Cooperative games with fuzzy coalitions are introduced in [4] and [5] to model situations where agents have the possibility to cooperate with different participation levels, varying from non-cooperation to full cooperation, and where the obtained reward depends on the levels of participation. A fuzzy coalition describes the participation levels at which each player is involved in cooperation. The use of fuzzy coalitions is especially appealing in joint projects where cooperating players possess some (divisible) private resources such as (divisible) commodities, time, money, and have to decide about the amount to be invested in the joint project. Another motivation for the use of fuzzy coalitions relies on a "large economy" issue: a fuzzy coalition in every finite economy becomes a crisp coalition in the related infinite non-atomic economy obtained by replacing each agent of the original finite economy by a continuum of identical agents. A cooperative fuzzy game with fuzzy coalitions (or fuzzy game) with a fixed set of players is represented by its characteristic function. The characteristic function of a fuzzy game specifies the worth of each fuzzy coalition as a real number. Since in classical cooperative games, which we henceforth refer to as crisp games, agents are either fully involved or not involved at all in cooperation with some other agents, one can look at the classical model of cooperative games as a rough (discrete) version of the model of cooperative fuzzy games.

Games with fuzzy coalitions and their applications have been receiving increased attention in the game theory literature. Basic monographs include [31] where attention is paid to triangular norm-based measures and special extensions of the diagonal Aumann-Shapley value (cf. [7]), and [14] where the stress in on noncooperative games. Fuzzy and multiobjective games are object of analysis also in [80]. Research in the theory of fuzzy games and their applications has also had as a result a significant number of scientific papers dealing with: solution concepts, their axiomatic characterizations and relations between them (cf. [17], [18], [19], [73], [76], [95], [114], [119], [120], [121]); special classes of fuzzy games and their interrelations (cf. [9], [17], [18], [19], [113], [114], [115], [119], [120], [121]); applications of fuzzy games in different situations (cf. [50], [66]); new interpretations of the model of a cooperative fuzzy game (cf. [10]).

We start this part with the definitions of a fuzzy coalition and a cooperative fuzzy game, and develop the theory of cooperative fuzzy games without paying extensively attention to the ways in which a game with crisp coalitions can be extended to a game with fuzzy coalitions. One possibility is to consider the multilinear extension of a crisp game introduced in [83] or to consider extensions that are given using the Choquet integral (cf. [35]). The reader who is interested in this extension problem is referred to [125] and [126].

This part is organized as follows. Chapter 6 introduces basic notation and notions from cooperative game theory with fuzzy coalitions. In Chapter 7 we have collected various set-valued and one-point solution concepts for fuzzy games like the Aubin core, the dominance core and stable sets, generalized cores and stable sets, as well as different core catchers and compromise values. Relations among these solution concepts are extensively studied. Chapter 8 is devoted to the notion of convexity of a cooperative fuzzy game. We present several characterizations of convex fuzzy games and study special properties of solution concepts. For this class of fuzzy games we also introduce and study the notion of participation monotonic allocation scheme and that of constrained egalitarian solution. In Chapter 9 we study the cone of fuzzy clan games together with related set-valued solution concepts for these games; the new solution concept of a bi-monotonic participation allocation scheme is introduced.

6

Preliminaries

Let N be a non-empty set of players usually of the form $\{1,\ldots,n\}$. From now on we systematically refer to elements of 2^N as *crisp coalitions*, and to cooperative games in G^N as *crisp games*.

Definition 6.1. *A* ***fuzzy coalition*** *is a vector* $s \in [0,1]^N$.

The i-th coordinate s_i of s is the *participation level* of player i in the fuzzy coalition s. Instead of $[0,1]^N$ we will also write $\mathcal{F}^N$ for the set of fuzzy coalitions on player set N.

A crisp coalition $S \in 2^N$ corresponds in a canonical way with the fuzzy coalition e^S, where $e^S \in \mathcal{F}^N$ is the vector with $\left(e^S\right)^i = 1$ if $i \in S$, and $\left(e^S\right)i = 0$ if $i \in N \setminus S$. The fuzzy coalition e^S corresponds to the situation where the players in S fully cooperate (i.e. with participation levels 1) and the players outside S are not involved at all (i.e. they have participation levels 0). In this part of the book we often refer to fuzzy coalitions e^S with $S \in 2^N$ as *crisp-like coalitions*. We denote by e^i the fuzzy coalition corresponding to the crisp coalition $S = \{i\}$ (and also the i-th standard basis vector in $\mathbb{R}^n$). The fuzzy coalition e^N is called the grand coalition, and the fuzzy coalition (the n-dimensional vector) $e^{\emptyset} = (0,\ldots,0)$ corresponds to the empty crisp coalition. We denote the set of all non-empty fuzzy coalitions by $\mathcal{F}_0^N = \mathcal{F}^N \setminus \{e^{\emptyset}\}$. Notice that we can identify the fuzzy coalitions with points in the hypercube $[0,1]^N$ and the crisp coalitions with the $2^{|N|}$ extreme points (vertices) of this hypercube. For $N = \{1,2\}$ we have a square with vertices $(0,0)$, $(0,1)$, $(1,0)$, $(1,1)$. The corresponding geometric picture for $N = \{1,2,3\}$ is that of a cube.

For $s \in \mathcal{F}^N$ we define the carrier of s by $car(s) = \{i \in N \mid s_i > 0\}$ and call s a *proper fuzzy coalition* if $car(s) \neq N$. The set of proper fuzzy coalitions on player set N is denoted by $\mathcal{PF}^N$, and the set of non-empty proper fuzzy coalitions on player set N by $\mathcal{PF}_0^N$.

For $s, t \in \mathcal{F}^N$ we use the notation $s \leq t$ iff $s_i \leq t_i$ for each $i \in N$. We define $s \wedge t = (\min(s_1, t_1), \ldots, \min(s_n, t_n))$ and $s \vee t = (\max(s_1, t_1), \ldots, \max(s_n, t_n))$. The set operations $\vee$ and $\wedge$ play the same role for the fuzzy coalitions as the union and intersection for crisp coalitions.

For $s \in \mathcal{F}^N$ and $t \in [0, 1]$, we set $\left(s^{-i} \parallel t\right)$ to be the element in $\mathcal{F}^N$ with $\left(s^{-i} \parallel t\right)_j = s_j$ for each $j \in N \setminus \{i\}$ and $\left(s^{-i} \parallel t\right)_i = t$.

For each $s \in \mathcal{F}^N$ we introduce the *degree of fuzziness* $\varphi(s)$ of s by $\varphi(s) = |\{i \in N \mid s_i \in (0, 1)\}|$. Note that $\varphi(s) = 0$ implies that s corresponds to a crisp coalition, and that in a coalition s with $\varphi(s) = n$ no participation level equals 0 or 1.

Definition 6.2. *A* ***cooperative fuzzy game*** *with player set N is a map $v : \mathcal{F}^N \to \mathbb{R}$ with the property $v(e^{\emptyset}) = 0$.*

The map v assigns to each fuzzy coalition a real number, telling what such a coalition can achieve in cooperation.

The set of fuzzy games with player set N will be denoted by FG^N. Note that FG^N is an infinite dimensional linear space.

Example 6.3. Let $v \in FG^{\{1,2,3\}}$ with

$$v(s_1, s_2, s_3) = \min\{s_1 + s_2, s_3\}$$

for each $s = (s_1, s_2, s_3) \in \mathcal{F}^{\{1,2,3\}}$. One can think of a situation where players 1, 2, and 3 have one unit of the infinitely divisible goods A, A, and B, respectively, where A and B are complementary goods, and where combining a fraction α of a unit of A and of B leads to a gain α.

Example 6.4. Let $v \in FG^{\{1,2\}}$ be defined by

$$v(s_1, s_2) = \begin{cases} 1 \text{ if } s_1 \geq \frac{1}{2},\ s_2 \geq \frac{1}{2}, \\ 0 \text{ otherwise,} \end{cases}$$

for each $s = (s_1, s_2) \in \mathcal{F}^{\{1,2\}}$. This game corresponds to a situation is which only coalitions with participation levels of the players of at least $\frac{1}{2}$ are winning, and all other coalitions are losing.

Example 6.5. (A public good game) Suppose n agents want to create a facility for joint use. The cost of the facility depends on the sum of the participation levels of the agents and it is described by $k\left(\sum_{i=1}^{n} s_i\right)$, where k is a continuous monotonic increasing function on $[0, n]$, with $k(0) = 0$, and where $s_1, \ldots, s_n \in [0,1]$ are the participation levels of the agents. The gain of an agent i with participation level s_i is given by $g_i(s_i)$, where the function $g_i : [0,1] \to \mathbb{R}$ is continuously monotonic increasing with $g_i(0) = 0$. This situation leads to a fuzzy game $v \in FG^N$ where $v(s) = \sum_{i=1}^n g_i(s_i) - k\left(\sum_{i=1}^n s_i\right)$ for each $s \in \mathcal{F}^N$.

For each $s \in \mathcal{F}^N$, let $\lceil s \rfloor := \sum_{i=1}^n s_i$ be the *aggregated participation level* of the players in N with respect to s. Given $v \in FG^N$ and $s \in \mathcal{F}_0^N$ we denote by $\alpha(s, v)$ the *average worth* of s with respect to $\lceil s \rfloor$, that is

$$\alpha(s, v) := \frac{v(s)}{\lceil s \rfloor}. \tag{6.1}$$

Note that $\alpha(s, v)$ can be interpreted as a per participation-level-unit value of coalition s.

As it is well known, the notion of a subgame plays an important role in the theory of cooperative crisp games. In what follows, the role of subgames of a crisp game will be taken over by restricted games of a fuzzy game.

For $s, t \in [0,1]^N$ let $s * t$ denote the *coordinate-wise product* of s and t, i.e. $(s * t)_i = s_i t_i$ for all $i \in N$.

Definition 6.6. *Let $v \in FG^N$ and $t \in \mathcal{F}_0^N$. The t-**restricted game** of v is the game $v_t : \mathcal{F}_0^N \to \mathbb{R}$ given by $v_t(s) = v(t * s)$ for all $s \in \mathcal{F}_0^N$.*

In a t-restricted game, $t \in \mathcal{F}_0^N$ plays the role of the grand coalition in the sense that the t-restricted game considers only the subset $\mathcal{F}_t^N$ of $\mathcal{F}_0^N$ consisting of fuzzy coalitions with participation levels of the corresponding players at most t, $\mathcal{F}_t^N = \{s \in \mathcal{F}_0^N \mid s \leq t\}$.

Remark 6.7. When $t = e^T$ then $v_t(s) = v(e^T * s) = v\left(\sum_{i \in T} s_i e^i\right)$ for each $s \in \mathcal{F}^N$, and for $s = e^S$ we obtain $v_t(e^S) = v(e^{S \cap T})$.

Special attention will be paid to fuzzy unanimity games. In the theory of cooperative crisp games unanimity games play an important role not only because they form a natural basis of the linear space G^N, but also since various interesting classes of games are nicely described with the aid of unanimity games.

For $t \in \mathcal{F}_0^N$, we denote by u_t the fuzzy game defined by

$$u_t(s) = \begin{cases} 1 \text{ if } s \geq t, \\ 0 \text{ otherwise.} \end{cases} \tag{6.2}$$

We call this game the *unanimity game based on t*: a fuzzy coalition s is winning if the participation levels of s exceed weakly the corresponding participation levels of t; otherwise the coalition is losing, i.e. it has value zero.

Remark 6.8. The game in Example 6.4 is the unanimity game with $t = \left(\frac{1}{2}, \frac{1}{2}\right)$.

Of course, the theory of cooperative crisp games has been an inspiration source for the development of the theory of cooperative fuzzy games. In the next chapters we will use operators from FG^N to G^N and from G^N to FG^N (cf. [83], [125], and [126]). In particular, we shall consider the multilinear operator $ml : G^N \to FG^N$ (cf. [83] and (3.9)) and the crisp operator $cr : FG^N \to G^N$. Here for a crisp game $v \in G^N$, *the multilinear extension* $ml(v) \in FG^N$ is defined by

$$ml(v)(s) = \sum_{S \in 2^N \setminus \{\emptyset\}} \left(\prod_{i \in S} s_i \prod_{i \in N \setminus S} (1 - s_i) \right) v(S) \text{ for each } s \in \mathcal{F}^N.$$

For a fuzzy game $v \in FG^N$, the corresponding *crisp game* $cr(v) \in G^N$ is given by

$$cr(v)(S) = v(e^S) \text{ for each } S \in 2^N.$$

Remark 6.9. In view of Remark 6.7, the restriction of $cr(v_{e^T}) : 2^N \to \mathbb{R}$ to 2^T is the subgame of $cr(v)$ on the player set T.

Example 6.10. For the crisp unanimity game u_T the multilinear extension is given by $ml\,(u_T)\,(s) = \prod_{i \in T} s_i$ (cf. [125] and [126]) and $cr\,(ml\,(u_T)) = u_T$. For the games $v, w \in FG^{\{1,2\}}$, where $v\,(s_1, s_2) = s_1\,(s_2)^2$ and $w\,(s_1, s_2) = s_1\sqrt{s_2}$ for each $s \in \mathcal{F}^{\{1,2\}}$, we have $cr\,(v) = cr\,(w)$.

In general the composition $cr \circ ml : G^N \to G^N$ is the identity map on G^N. But $ml \circ cr : FG^N \to FG^N$ is not the identity map on FG^N if $|N| \geq 2$.

Example 6.11. Let $N = \{1, 2\}$ and let $v \in FG^{\{1,2\}}$ be given by $v(s_1, s_2) = 5 \min\{s_1, 2s_2\}$ for each $s = (s_1, s_2) \in \mathcal{F}^{\{1,2\}}$. Then for the crisp game $cr(v)$ we have $cr(v)(\{1\}) = cr(v)(\{2\}) = 0$ and $cr(v)(\{1, 2\}) = 5$. For the multilinear extension $ml(cr(v))$ we have $ml(cr(v)\,(s)) = 5s_1s_2$. Notice that $ml\left(cr\,(v)\left(1, \frac{1}{2}\right)\right) = 2\frac{1}{2}$ but $v\left(1, \frac{1}{2}\right) = 5$ implying that $v \neq ml(cr(v))$.

Remark 6.12. Note that for a unanimity game u_t, the corresponding crisp game $cr(u_t)$ is equal to u_T, where u_T is the crisp unanimity game based on $T = car(t)$. Conversely, $ml(u_T)$ is for no $T \in 2^N \setminus \{\emptyset\}$ a fuzzy unanimity game because $ml(u_T)$ has a continuum of values: $ml(u_T)(s) = \prod_{i \in T} s_i$ for each $s \in \mathcal{F}^N$.

7

Solution Concepts for Fuzzy Games

In this chapter we introduce several solution concepts for fuzzy games and study their properties and interrelations. Sections 7.1-7.3 are devoted to various core concepts and stable sets. The Aubin core introduced in Section 7.1 plays a key role in the rest of this chapter. Section 7.4 presents the Shapley value and the Weber set for fuzzy games which are based on crisp cooperation and serve as an inspiration source for the path solutions and the path solution cover introduced in Section 7.5. Compromise values for fuzzy games are introduced and studied in Section 7.6.

7.1 Imputations and the Aubin Core

Definition 7.1. *The* ***imputation set*** $I(v)$ *of* $v \in FG^N$ *is the set*

$$\left\{x \in \mathbb{R}^n \mid \sum_{i\in N} x_i = v(e^N),\ x_i \geq v(e^i) \text{ for each } i \in N\right\}.$$

Definition 7.2. *The* ***Aubin core*** *(cf. [4], [5], and [6])* $C(v)$ *of a fuzzy game* $v \in FG^N$ *is the set*

$$C(v) = \left\{x \in I(v) \mid \sum_{i\in N} s_i x_i \geq v(s) \text{ for each } s \in \mathcal{F}^N\right\}.$$

So, $x \in C(v)$ can be seen as a distribution of the value of the grand coalition e^N, where for each fuzzy coalition s the total payoff is not smaller than $v(s)$; thus, $C(v)$ is the subset of imputations which are stable against any possible deviation by fuzzy coalitions. Notice that each player $i \in N$ is supposed to gain a payoff proportional to his participation level when he partially cooperates. That is, if player i gains x_i for full cooperation then his gain will be $s_i x_i$ in case he participates at level s_i.

Note that the Aubin core $C(v)$ of $v \in FG^N$ can be also defined as $\left\{x \in I(v) \mid \sum_{i \in N} s_i x_i \geq v(s) \text{ for each } s \in \mathcal{F}_0^N\right\}$ and this is more in the spirit of the definition of the core of a crisp game (cf. Definition 2.3).

Remark 7.3. The core $C(cr(v))$ of the crisp game corresponding to v includes $C(v)$: $C(v) \subset C(cr(v))$. We will see in the next chapter that for convex fuzzy games the two cores coincide.

Clearly, the Aubin core $C(v)$ of a fuzzy game v is a closed convex subset of $\mathbb{R}^n$ for each $v \in FG^N$. Of course, the Aubin core may be empty as Example 7.4 shows or can consist of a single element as in Example 7.5.

Example 7.4. Consider again the game v in Example 6.4. The core $C(v)$ is empty because for a core element x it should hold $x_1 + x_2 = v\left(e^{\{1,2\}}\right) = 1$ and also $\frac{1}{2}x_1 + \frac{1}{2}x_2 \geq v\left(\frac{1}{2}, \frac{1}{2}\right) = 1$, which is impossible.

Example 7.5. For the game in Example 6.3, good B is scarce in the grand coalition which is reflected in the fact that the core consists of one point $(0,0,1)$, corresponding to the situation where all gains go to player 3 who possesses the scarce good.

The next proposition shows that for a unanimity game u_t (cf. (6.2)) every arbitrary division of 1 among players who have participation level 1 in t generates a core element (cf. [17]).

Proposition 7.6. *Let $u_t \in FG^N$ be the unanimity game based on the fuzzy coalition $t \in \mathcal{F}_0^N$. Then the Aubin core $C(u_t)$ is non-empty iff $t_k = 1$ for some $k \in N$. In fact*

$$C(u_t) = co\left\{e^k \mid k \in N, t_k = 1\right\}.$$

Proof. If $t_k = 1$ for some $k \in N$, then $e^k \in C(u_t)$. Therefore, $co\left\{e^k \mid k \in N, t_k = 1\right\} \subset C(u_t)$.

Conversely, $x \in C(u_t)$ implies that $\sum_{i=1}^n x_i = 1 = u_t(e^N)$, $\sum_{i=1}^n t_i x_i \geq 1 = u_t(t)$, $x_i \geq u_t(e^i) \geq 0$ for each $i \in N$. So $x \geq 0$, $\sum_{i=1}^n x_i(1-t_i) \leq 0$, which implies that $x_i(1-t_i) = 0$ for all $i \in N$. Hence, $\{x_i \mid x_i > 0\} \subset \{i \in N \mid t_i = 1\}$ and, consequently, $x \in co\left\{e^k \mid k \in N, t_k = 1\right\}$. So, $C(u_t) \subset co\left\{e^k \mid k \in N, t_k = 1\right\}$.

In what follows, we denote by FG_*^N the set of fuzzy games with a non-empty (Aubin) core.

7.2 Cores and Stable Sets

Now, we introduce two other cores for a fuzzy game $v \in FG^N$, namely the proper core and the crisp core, by weakening the stability conditions in the definition of the Aubin core (cf. [114]).

Definition 7.7. *The* **proper core** $C^P(v)$ *of* $v \in FG^N$ *is the set*

$$C^P(v) = \left\{ x \in I(v) \mid \sum_{i \in N} s_i x_i \geq v(s) \textit{ for each } s \in \mathcal{PF}^N \right\}.$$

Note that in the definition of $C^P(v)$ only stability regarding proper fuzzy coalitions is considered (cf. Chapter 6, page 78). The set $C^P(v)$ can be also defined as

$$\left\{ x \in I(v) \mid \sum_{i \in N} s_i x_i \geq v(s) \text{ for each } s \in \mathcal{PF}_0^N \right\}.$$

Further, we can consider only crisp-like coalitions e^S in the stability conditions.

Definition 7.8. *The* **crisp core** $C^{cr}(v)$ *of* $v \in FG^N$ *is the set*

$$C^{cr}(v) = \left\{ x \in I(v) \mid \sum_{i \in S} x_i \geq v(e^S) \textit{ for each } S \in 2^N \right\}.$$

Clearly, the crisp core $C^{cr}(v)$ of a fuzzy game v can be also defined as $\left\{ x \in I(v) \mid \sum_{i \in S} x_i \geq v(e^S) \text{ for each } S \in 2^N \setminus \{\emptyset\} \right\}$ and it is also the core of the crisp game $cr(v)$. One can easily see that both cores $C^P(v)$ and $C^{cr}(v)$ are convex sets.

Let $v \in FG^N$, $x, y \in I(v)$ and let $s \in \mathcal{F}_0^N$. We say that x *dominates* y *via* s, denoted by $x \operatorname{dom}_s y$, if

(i) $x_i > y_i$ for all $i \in car(s)$, and

(ii) $\sum_{i \in N} s_i x_i \leq v(s)$.

The two above conditions are interpreted as follows:

- $x_i > y_i$ implies $s_i x_i > s_i y_i$ for each $i \in car(s)$, which means that the imputation $x = (x_1, ..., x_n)$ is better than the imputation $y = (y_1, ..., y_n)$ for all (active) players $i \in car(s)$;
- $\sum_{i \in N} s_i x_i \leq v(s)$ means that the payoff $\sum_{i \in N} s_i x_i$ is reachable by the fuzzy coalition s.

Remark 7.9. Note that $x\,\mathrm{dom}_s\,y$ implies $s \in \mathcal{PF}_0^N$ because from $x_i > y_i$ for all $i \in N$ it follows $\sum_{i\in N} x_i > \sum_{i\in N} y_i$, in contradiction with $x, y \in I(v)$. It is, however, to be noted that $|car(s)| = 1$ is possible.

We simply say x *dominates* y, denoted by $x\,\mathrm{dom}\,y$, if there is a nonempty (proper) fuzzy coalition s such that $x\,\mathrm{dom}_s\,y$. The negation of $x\,\mathrm{dom}\,y$ is denoted by $\neg\, x\,\mathrm{dom}\,y$.

Definition 7.10. *The* ***dominance core*** *$DC(v)$ of a fuzzy game $v \in FG^N$ is the set of imputations which are not dominated by any other imputation,*

$$DC(v) = \{x \in I(v) \mid \neg\, y\,\mathrm{dom}\,x \textit{ for all } y \in I(v)\}.$$

Definition 7.11. *A* ***stable set*** *of a fuzzy game $v \in FG^N$ is a nonempty set K of imputations satisfying the properties:*
(i) (Internal stability) For all $x, y \in K$, $\neg\, x\,\mathrm{dom}\,y$;
(ii) (External stability) For all $z \in I(v) \setminus K$, there is an imputation $x \in K$ such that $x\,\mathrm{dom}\,z$.

Theorem 7.12. *Let $v \in FG^N$. Then*
(i) $C(v) \subset C^P(v) \subset C^{cr}(v)$;
(ii) $C^P(v) \subset DC(v)$;
(iii) for each stable set K it holds that $DC(v) \subset K$.

Proof. The theorem is trivially true if $I(v) = \emptyset$. So, suppose in the following that $I(v) \neq \emptyset$.

(i) This follows straightforwardly from the definitions.

(ii) Let $x \in I(v) \setminus DC(v)$. Then there are $y \in I(v)$ and $s \in \mathcal{PF}_0^N$ satisfying $y_i > x_i$ for each $i \in car(s)$ and $\sum_{i\in N} s_i y_i \leq v(s)$. Then $\sum_{i\in car(s)} s_i x_i < \sum_{i\in car(s)} s_i y_i \leq v(s)$. Hence $x \in I(v) \setminus C^P(v)$. We conclude that $C^P(v) \subset DC(v)$.

(iii) Let K be a stable set. Since $DC(v)$ consists of undominated imputations and each imputation in $I(v) \setminus K$ is dominated by some imputation by the external stability property, it follows that $DC(v) \subset K$.

In the next theorem we give sufficient conditions for the coincidence of the proper core and the dominance core for fuzzy games.

Theorem 7.13. *Let $v \in FG^N$. Suppose $v\left(e^N\right) - \sum_{i\in N\setminus car(s)} v\left(e^i\right) - \frac{v(s)}{s^*} \geq 0$ for each $s \in \mathcal{F}_0^N$, where $s^* = \min_{i\in car(s)} s_i$. Then $C^P(v) = DC(v)$.*

Proof. Note that $C^P(v) = DC(v) = \emptyset$ if $I(v) = \emptyset$. Suppose $I(v) \neq \emptyset$. From Theorem 7.12 it follows that $C^P(v) \subset DC(v)$. We show the converse inclusion by proving that $x \notin C^P(v)$ implies $x \notin DC(v)$. Let $x \in I(v) \setminus C^P(v)$. Then there is $s \in \mathcal{PF}_0^N$ such that $\sum_{i\in N} s_i x_i < v(s)$. For each $i \in car(s)$ take $\varepsilon_i > 0$ such that $\sum_{i\in car(s)} s_i(x_i + \varepsilon_i) = v(s)$. Since

$$\sum_{i\in car(s)} (x_i + \varepsilon_i) \leq \sum_{i\in car(s)} \frac{s_i\,(x_i + \varepsilon_i)}{s^*} = \frac{v(s)}{s^*},$$

we can take $\delta_i \geq 0$ for each $i \notin car(s)$ such that $\sum_{i\notin car(s)} \delta_i = \frac{v(s)}{s^*} - \sum_{i\in car(s)} (x_i + \varepsilon_i)$. Define $y \in \mathbb{R}^N$ by

$$y_i = \begin{cases} x_i + \varepsilon_i & \text{for each } i \in car(s), \\ v\left(e^i\right) + \frac{v(e^N) - \sum_{i\in N\setminus car(s)} v(e^i) - \frac{v(s)}{s^*}}{|N\setminus car(s)|} + \delta_i & \text{for each } i \notin car(s). \end{cases}$$

Note that $\sum_{i\in N} y_i = v(e^N)$, $y_i > x_i > v(e^i)$ for each $i \in car(s)$ and, since $v(e^N) - \sum_{i\in N\setminus car(s)} v(e^i) - \frac{v(s)}{s^*} \geq 0$, we have $y_i \geq v(e^i)$ for each $i \in N \setminus car(s)$. Hence $y \in I(v)$. Now, since $y_i > x_i$ for all $i \in car(s)$ and $\sum_{i\in N} s_i y_i = v(s)$ we have $y \,\mathrm{dom}_s\, x$; thus $x \in I(v) \setminus DC(v)$.

Remark 7.14. Let $v \in FG^N$. Take the crisp game $w = cr(v)$. Then $v(e^N) \geq \frac{v(s)}{s^*} + \sum_{i\in N\setminus car(s)} v(e^i)$ for each $s \in \mathcal{F}_0^N$ implies $w(N) \geq w(S) + \sum_{i\in N\setminus S} w(i)$, for each $S \subseteq N$. So, Theorem 7.13 can be seen as an extension of the corresponding property for cooperative crisp games (cf. Theorem 2.13(i)).

Now, we give two examples to illustrate the results in the above theorems.

Example 7.15. Let $N = \{1, 2\}$ and let $v : \mathcal{F}^{\{1,2\}} \to \mathbb{R}$ be given by $v(s_1, s_2) = s_1 + s_2 - 1$ for each $s \in \mathcal{F}_0^{\{1,2\}}$ and $v(e^{\emptyset}) = 0$. Further, let

$$v_1(s) = \begin{cases} v(s) & \text{if } s \neq \left(0, \frac{1}{2}\right), \\ 4 & \text{if } s = \left(0, \frac{1}{2}\right). \end{cases}$$

$$v_2(s) = \begin{cases} v(s) & \text{if } s \neq \left(\frac{1}{2}, \frac{1}{2}\right), \\ 4 & \text{if } s = \left(\frac{1}{2}, \frac{1}{2}\right). \end{cases}$$

Let $\Delta = \{x \in \mathbb{R}^2 \mid x_1 \geq 0,\ x_2 \geq 0,\ x_1 + x_2 = 1\}$. Then
(i) $C(v) = C^P(v) = DC(v) = I(v) = \Delta$,
(ii) $C(v_1) = C^P(v_1) = \emptyset,\ DC(v_1) = I(v_1) = \Delta$,

(iii) $C(v_2) = \emptyset$, $C^P(v_2) = DC(v_2) = I(v_2) = \Delta$,
(iv) for v, v_1, and v_2, the imputation set Δ is the unique stable set.
Note that $2v_2(s) + \sum_{i \in N \setminus car(s)} v_2(e^i) > v_2(e^N)$ for $s = \left(\frac{1}{2}, \frac{1}{2}\right)$ and $C^P(v_2) = DC(v_2)$. Hence, the sufficient condition in Theorem 7.13 for the equality $C^P(v) = DC(v)$ is not a necessary condition.

In the next example we give a fuzzy game v with $C(v) \neq DC(v)$ and $C(v) \neq \emptyset$. Notice that for a crisp game w we have that $C(w) = \emptyset$ if $C(w) \neq DC(w)$ (cf. Theorem 2.13(ii)).

Example 7.16. Let $N = \{1, 2\}$ and let $v : \mathcal{F}^{\{1,2\}} \to \mathbb{R}$ be given by $v(s_1, 1) = \sqrt{s_1}$ for all $(s_1, 1) \in \mathcal{F}^{\{1,2\}}$, and $v(s_1, s_2) = 0$ otherwise. Then $I(v) = \{(x_1, x_2) \in \mathbb{R}^2_+ \mid x_1 + x_2 = 1\}$, $C(v) = \{x \in I(v) \mid 0 \leq x_1 \leq \frac{1}{2}\}$ $\neq I(v)$, and $C^P(v) = DC(v) = I(v)$. Further, $I(v)$ is the unique stable set.

By using the average worth of a coalition $s \in \mathcal{F}_0^N$ in a game $v \in FG^N$ (cf. (6.1)), we define another core concept for fuzzy games.

Definition 7.17. *The* ***equal division core*** *$EDC(v)$ of $v \in FG^N$ is the set*

$$\left\{x \in I(v) \mid \nexists s \in \mathcal{F}_0^N \text{ s.t. } \alpha(s, v) > x_i \text{ for all } i \in car(s)\right\}.$$

So $x \in EDC(v)$ can be seen as a distribution of the value of the grand coalition e^N, where for each fuzzy coalition s, there is a player i with a positive participation level for which the payoff x_i is at least as good as the equal division share $\alpha(s, v)$ of $v(s)$ in s.

Proposition 7.18. *Let $v \in FG^N$. Then*
(i) $EDC(v) \subset EDC(cr(v))$;
(ii) $C(v) \subset EDC(cr(v))$.

Proof. (i) Suppose $x \in EDC(v)$. Then by the definition of $EDC(v)$ there is no $e^S \in \mathcal{F}_0^N$ s.t. $\alpha\left(e^S, v\right) > x_i$ for all $i \in car(e^S)$. Taking into account that $cr(v)(S) = v\left(e^S\right)$ for all $S \in 2^N$, there is no $S \neq \emptyset$ s.t. $\frac{cr(v)(S)}{|S|} > x_i$ for all $i \in S$. Hence, $x \in EDC(cr(v))$.

(ii) Suppose $x \notin EDC(v)$. Then there exists $s \in \mathcal{F}_0^N$ s.t. $\alpha(s, v) > x_i$ for all $i \in car(s)$. Then

$$\sum_{i=1}^{n} s_i x_i < \sum_{i=1}^{n} \alpha(s, v)\, s_i = v(s)$$

which implies that $x \notin C(v)$. So $C(v) \subset EDC(v)$.

The next example shows that $EDC(v)$ and $EDC(cr(v))$ are not necessarily equal.

Example 7.19. Let $N = \{1, 2, 3\}$ and $v(s_1, s_2, s_3) = \sqrt{s_1 + s_2 + s_3}$ for each $s = (s_1, s_2, s_3) \in \mathcal{F}^{\{1,2,3\}}$. For this game we have

$$EDC(cr(v)) = \left\{ \frac{\sqrt{3}}{3}, \frac{\sqrt{3}}{3}, \frac{\sqrt{3}}{3} \right\} \text{ and } EDC(v) = \emptyset.$$

7.3 Generalized Cores and Stable Sets

As investigated in [76], core elements (à la Aubin) of a fuzzy game can be associated with additive separable supporting functions of fuzzy games. Thus, these authors introduced generalized cores whose elements consist of more general separable supporting functions of the game and studied their properties and relations to stable sets.

The main results on cores and stable sets of a fuzzy game from the previous sections have been extended based on some families of separable supporting functions for fuzzy games which are *monotonic*, i.e., for any $s, s' \in \mathcal{F}^N$, if $s_i \leq s_i'$ for all $i \in N$ then $v(s) \leq v(s')$. Since $v(e^{\emptyset}) = 0$, the monotonicity of v implies that $v(s) \geq 0$ for all $s \in \mathcal{F}^N$. We assume throughout this section that v is monotonic.

Looking with a geometrical eye to the Aubin core, we note first that to each $x \in \mathbb{R}^n$ one can associate in a canonical way a linear function $x^* : \mathbb{R}^n \to \mathbb{R}$, assigning to each $s \in \mathbb{R}^n$ the real number $x^*(s) = \sum_{i \in N} s_i x_i$. For a fuzzy game $v \in FG^N$ we have then $x \in C(v)$ if and only if $x^*(e^N) = v(e^N)$ and $x^*(s) \geq v(s)$ for each $s \in \mathcal{F}^N$. In geometric terms: $x \in C(v)$ if and only if $graph(x^*) = \{(s, \alpha) \in \mathbb{R}^n \times \mathbb{R} \mid \alpha = x^*(s)\}$ lies 'above' $graph(v) = \{(s, \alpha) \in \mathcal{F}^N \times \mathbb{R} \mid \alpha = v(s)\}$ and the two graphs touch each other at the points $(e^{\emptyset}, v(e^{\emptyset})), (e^N, v(e^N)) \in \mathbb{R}^n \times \mathbb{R}$. Since x^* is a linear function, $graph(x^*)$ is a supporting hyperplane at $(e^N, v(e^N))$ of $graph(v)$ in case $x \neq 0$. Note further that $x^* : \mathbb{R}^n \to \mathbb{R}$ given above is a special class of the separable form $\sum_{i=1}^n p(s_i) x_i$ for $s \in \mathcal{F}^N$, where $p : [0, 1] \to [0, 1]$ with $p(0) = 0$ and $p(1) = 1$, with $p(s_i) = s_i \in [0, 1]$ for each $i \in N$.

In the following we will consider also non-linear functions $p : [0, 1] \to [0, 1]$ with $p(0) = 0$ and $p(1) = 1$. The function p is called monotonic if for any $a, b \in [0, 1]$ with $a \leq b$, $p(a) \leq p(b)$ holds. Let $\mathbf{P} = \{p : [0, 1] \to [0, 1] \mid p \text{ is monotonic and } p(0) = 0,\ p(1) = 1\}$. The function p is a *payment scheme* in case of partial cooperation. That is, if a player $i \in N$ gains x_i in case of full cooperation, then he gains $p(s_i) x_i$ when he participates at level s_i.

Definition 7.20. *Given a payment scheme* $p \in \mathbf{P}$*, the core based on* p *of a fuzzy game* v*, henceforth called the* **p-core** *of* v*, is given by*

$$C_p(v) = \left\{ x \in \mathbb{R}^n \mid \sum_{i \in N} x_i = v(e^N), \sum_{i \in N} p(s_i) x_i \geq v(s) \text{ for each } s \in \mathcal{F}^N \right\}.$$

The Aubin core is a special case of the p-core, where p is given by $p(s_i) = s_i$ for each $i \in N$. While the Aubin core assumes only a payment proportional to players' participation level, the p-core allows much wider payment schemes for partial participation.

The following example may help for a better understanding of the notion of the p-core. Suppose that full participation for a worker means that he works for thirty days and his total reward is \$3,000. The Aubin core assumes that a portion of \$3,000 is paid in proportion to the worker's participation level. A typical example is a daily wage of \$100. However, other payment schemes are common practice in the economies as well. For instance, the full amount of \$3,000 can be paid at the beginning, or half of this amount is paid at the beginning and the rest after the job is completed the job, and so on. This motivates the study of general payment rules.

The two extreme payment rules, i.e., the one which gives the whole amount to a player when he starts participating, and the other which gives the whole amount after full cooperation is reached, are denoted by p^+ and p^-, respectively:

$$p^+(a) = \begin{cases} 1 & \text{for } 0 < a \leq 1, \\ 0 & \text{for } a = 0; \end{cases}$$

$$p^-(a) = \begin{cases} 1 & \text{for } a = 1, \\ 0 & \text{for } 0 \leq a < 1. \end{cases}$$

In an obvious way one can define the *proper p-core* of $v \in FG^N$, i.e.,

$$C_p^P(v) = \left\{ x \in \mathbb{R}^n \mid \sum_{i \in N} x_i = v(e^N), \sum_{i \in N} p(s_i) x_i \geq v(s) \text{ for each } s \in P\mathcal{F}^N \right\}$$

and the *crisp p-core* of $v \in FG^N$, i.e.,

$$C_p^{cr}(v) = \left\{ x \in \mathbb{R}^n \mid \sum_{i \in N} x_i = v(e^N), \sum_{i \in S} p(1) x_i \geq v(e^S) \text{ for each } S \in 2^N \right\}.$$

Since $p(1) = 1$, the crisp p-core is the same as the crisp core $C^{cr}(v)$.

Remark 7.21. It is easily noticed that $C_p(v)$, $C_p^P(v)$, and $C_p^{cr}(v)$ $(= C^{cr}(v))$ are all convex sets.

For the p-core and the proper p-core, the following properties hold.

Theorem 7.22. *Let $v \in FG^N$. Then,*
(i) For any $p, p' \in \mathbf{P}$, if $p(a) \le p'(a)$ for all $a \in [0,1]$, then $C_p(v) \subseteq C_{p'}(v)$ and $C_p^P(v) \subseteq C_{p'}^P(v)$. Thus, $C_{p^-}(v) \subseteq C_p(v) \subseteq C_{p^+}(v)$ and $C_{p^-}^P(v) \subseteq C_p^P(v) \subseteq C_{p^+}^P(v)$ for all $p \in \mathbf{P}$;
(ii) For any $p \in \mathbf{P}$, $C_p(v) \subseteq C_p^P(v) \subseteq C^{cr}(v)$;
(iii) $C_{p^+}(v) = C_{p^+}^P(v) = C^{cr}(v)$.

Proof. (i) Take any $x \in C_p(v)$. Then $\sum_{i\in N} x_i = v(e^N)$ and $\sum_{i\in N} p(s_i)x_i \ge v(s)$ for all $s \in \mathcal{F}^N$. Since $x \in I(v)$ implies $x_i \ge v(e^i) \ge 0$, if $p(a) \le p'(a)$ for all $a \in [0,1]$, then $p(s_i)x_i \le p'(s_i)x_i$ for all $i \in N$. Hence, $\sum_{i\in N} p'(s_i)x_i \ge v(s)$ for all $s \in \mathcal{F}^N$, which implies $x \in C_{p'}(v)$. A similar proof applies to $C_p^P(v) \subseteq C_{p'}^P(v)$.

(ii) It is clear from the definitions of $C_p(v)$, $C_p^P(v)$ and $C^{cr}(v)$.

(iii) $C_{p^+}(v) \subseteq C^{cr}(v)$ follows from (ii). To show the reverse inclusion, take any $x \in C^{cr}(v)$. Then $\sum_{i\in N} x_i = v(e^N)$ and $\sum_{i\in S} x_i \ge v(e^S)$ for all $S \subseteq N$. Take any $s \in \mathcal{F}^N$. It suffices to show that $\sum_{i\in car(p^+(s))} p^+(s_i)x_i \ge v(s)$. From the definition of p^+, $p^+(s_i) = 1$ if $s_i > 0$ and $p^+(s_i) = 0$ if $s_i = 0$. Thus, we obtain $\sum_{i\in car(p^+(s))} p^+(s_i)x_i = \sum_{i:s_i>0} x_i$. Let $S = \{i \mid s_i > 0\}$. Then $\sum_{i\in car(p(s))} p(s_i)x_i = \sum_{i:s_i>0} x_i = \sum_{i\in S} x_i \ge v(e^S) \ge v(s)$. The last inequality follows from the monotonicity of v.

Furthermore, the p-core of a unanimity game is nonempty when $p = p^+$. However, the Aubin core may be empty for unanimity games, as shown in Example 7.4 (see also Proposition 7.6). In fact, we have the following theorem.

Theorem 7.23. *Let u_t be a unanimity game on the player set N. Then $C_{p^+}(u_t) \ne \emptyset$.*

Proof. Take $x \in I(u_t)$, that is $x \in \mathbb{R}^n$ where $\sum_{i=1}^n x_i = u_t(e^N) = 1$ and $x_i \ge u_t(e^i) = 0$ for $i \in N$. For p^+, we have then $x \in C_{p^+}(u_t)$. Take any fuzzy coalition $s \in \mathcal{F}^N$. The following two situations may appear:

- If $s_i > t_i$ for all $i \in N$, then $u_t(s) = 1$. From the definition of p^+, we have $p(s_i) = 1$ for all $i \in N$ since $s_i \ge t_i$ and $t_i > 0$ for all $i \in N$. Hence, $\sum_{i=1}^n p^+(s_i)x_i = \sum_{i=1}^n x_i = 1 = u_t(s)$.

– If $s_i < t_i$ for some $i \in N$, then $u_t(s) = 0$. Since $p^+(a) \geq 0$ for all $a \in [0,1]$ and $x_i \geq 0$, we have $\sum_{i=1}^n p^+(s_i)x_i \geq 0 = u_t(s)$.

Remark 7.24. There exist various $p \in \mathbf{P}$ other than p^+ satisfying $C_p(u_t) \neq \emptyset$. An example is $p \in \mathbf{P}$ given by

$$p(a) = \begin{cases} \frac{1}{t^*}a & \text{if } a \in [0, t^*], \\ 1 & \text{if } a \in (t^*, 1], \end{cases}$$

where $t^* = \min(t_1, \ldots, t_n)$. In a similar manner to the proof above, one may show that $C_p(u_t)$ is not empty.

Now, we define generalized domination relations. Take a fuzzy game $v \in FG^N$ and a function $p \in \mathbf{P}$. For any two imputations $x, y \in I(v)$ and any fuzzy coalition $s \in \mathcal{F}^N$, we say that x p-dominates y via s, and denote it by $x \operatorname{dom}_s^p y$, if $p(s_i)x_i > p(s_i)y_i$ for all $i \in car(p(s))$, and $\sum_{i \in N} p(s_i)x_i \leq v(s)$. If there is at least one $s \in \mathcal{F}^N$ with $x \operatorname{dom}_s^p y$, we simply say x p-dominates y and denote it by $x \operatorname{dom}^p y$. Similarly to the case of $x \operatorname{dom}_s y$, we must have $|car(s)| < n$ if $x \operatorname{dom}_s^p y$.

Definition 7.25. *The* ***p-dominance core*** $DC_p(v)$ *of a fuzzy game* v *is a set of imputations which are not p-dominated by any other imputation.*

Definition 7.26. *A* ***p-stable set*** K_p *of a fuzzy game* v *is a set of imputations satisfying the following properties:*
(i) (p-internal stability) For all $x, y \in K_p$, $\neg x \operatorname{dom}^p y$;
(ii) (p-external stability) For all $z \in I(v) \setminus K_p$, *there is an imputation* $x \in K_p$ *such that* $x \operatorname{dom}^p z$.

Then, trivially, we have the extension of Theorem 7.12 (ii) and (iii).

Theorem 7.27. *Let* $v \in FG^N$ *and* $p \in \mathbf{P}$. *Then,*
(i) $C_p^P(v) \subseteq DC_p(v)$;
(ii) For each stable set $K_p : DC_p(v) \subseteq K_p$;
(iii) $x \operatorname{dom}^{p^+} y$ *in* v *if and only if* $x \operatorname{dom} y$ *in* $w = cr(v)$. *Thus,* $DC_{p^+}(v) = DC(w)$ *and* K *is a stable set in* v *under* p^- *if and only if* K *is a stable set in* w.

Proof. (i) The theorem is trivially true if $DC_p(v) = I(v)$. So, suppose $DC_p(v) \subset I(v)$. Let $x \in I(v) \setminus DC_p(v)$. Then there are $y \in I(v)$ and $s \in \mathcal{PF}^N$ satisfying $p(s_i)y_i > p(s_i)x_i$ for each $i \in car(p(s))$ and $\sum_{i \in car(p(s))} p(s_i)y_i \leq v(s)$. Then $\sum_{i \in car(p(s))} p_i(s_i)x_i < \sum_{i \in car(p(s))} p(s_i)y_i \leq v(s)$. Hence, $x \in I(v) \setminus C_p^P(v)$. We conclude that $C_p^P(v) \subset DC_p(v)$.

(ii) Let K_p be a stable set. Since $DC_p(v)$ consists of undominated imputations and each imputation in $I(v) \setminus K_p$ is dominated by some imputation by the external stability property, it follows that $DC_p(v) \subset K_p$.

(iii) Suppose $x \text{ dom}_s^{p^+} y$ in v. Then $\sum_{i \in car(p(s))} p^+(s_i)x_i \leq v(s)$ and $p^+(s_i)x_i > p^+(s_i)y_i$ for all $i \in car(p(s))$. Since $p^+(s_i) = 1$ if and only if $s_i > 0$, from the monotonicity of v we obtain $\sum_{i \in car(p(s))} x_i \leq v(e^{car(p(s))})$ and $x_i > y_i$ for all $i \in car(p(s))$. Thus, x dom y in w. The reverse direction holds true, since by $x \text{ dom}_S y$ in w we have $x \text{ dom}_{e^S}^{p^+} y$ in v. Then the latter half of (iii) easily follows.

In the next theorem we give a sufficient condition for the coincidence of the proper p-core and the p-dominance core.

Theorem 7.28. *Let $v \in FG^N$ and $p \in \mathbf{P}$ with $p(a) > 0$ for all $a \in (0,1]$. For each $s \in \mathcal{F}^N$ with $car(p(s)) \neq \emptyset$, let $p^*(s) = \min_{i \in car(p(s))} p(s_i)$ and $v_p^*(s) = \frac{v(s)}{p^*(s)}$. Suppose*

$$v(e^N) - v_p^*(s) - \sum_{i \in N \setminus car(p(s))} v(e^i) \geq 0$$

for each $s \in \mathcal{PF}^N$. Then $C_p^P(v) = DC_p(v)$, and thus $DC_p(v)$ is a convex set.

Proof. By Theorem 7.27(i), $C_p^P(v) \subseteq DC_p(v)$. We show the converse inclusion by proving that $x \notin C_p^P(v)$ implies $x \notin DC_p(v)$. If $I(v) = C_p^P(v)$, then we easily have $C_p^P(v) = DC_p(v)$ since $C_p^P(v) \subseteq DC_p(v) \subseteq I(v)$. We assume now $C_p^P(v) \subset I(v)$ and take $x \in I(v) \setminus C_p^P(v)$. Then there is $s \in \mathcal{PF}^N$ such that $\sum_{i \in car(p(s))} p(s_i)x_i < v(s)$. Then $\sum_{i \in car(p(s))} p^*(s)x_i < v(s)$, and thus $\sum_{i \in car(p(s))} x_i < v_p^*(s)$. Hence, for each $i \in car(p(s))$, we can take $\epsilon_i > 0$ such that $\sum_{i \in car(p(s))}(x_i + \epsilon_i) < v_p^*(s)$ and $\sum_{i \in car(p(s))} p(s_i)(x_i + \epsilon_i) < v(s)$. Define $y \in \mathbb{R}^n$ by

$$y_i = \begin{cases} x_i + \epsilon_i & \text{if } i \in car(p(s)), \\ v(e^i) + \frac{v(e^N) - v_p^*(s) - \sum_{j \in N \setminus car(p(s))} v(e^j)}{|N \setminus car(p(s))|} + \delta_i & \text{if } i \in N \setminus car(p(s)), \end{cases}$$

where $\delta_i > 0$ for all $i \in N \setminus car(p(s))$ are such that $\sum_{i \in N} y_i = v(e^N)$. Since $\sum_{i \in car(p(s))}(x_i + \epsilon_i) < v_p^*(s)$, we can take such δ_i, $i \in N \setminus car(p(s))$. Note that $y_i > x_i \geq v(e^i)$ for each $i \in car(p(s))$. Furthermore, since $v(e^N) - v_p^*(s) - \sum_{i \in \setminus car(p(s))} v(e^i) \geq 0$, we have $y_i > v(e^i)$ for each

$i \in N \setminus car(p(s))$. Hence, $y \in I(v)$. Now, since $y_i > x_i$ for all $i \in car(p(s))$, we have $p(s_i)y_i > p(s_i)x_i$ for all $i \in car(p(s))$. Moreover, $\sum_{i \in car(p(s))} p(s_i)y_i < v(s)$, and thus $y \text{ dom}^p x$. Hence, we obtain $x \in I(v) \setminus DC^p(v)$. Finally, by Remark 7.21, $DC^p(v)$ is a convex set.

7.4 The Shapley Value and the Weber Set

Let $\pi(N)$ be the set of linear orderings of N. We introduce for $v \in FG^N$ the *marginal vectors* $m^\sigma(v)$ for each $\sigma \in \pi(N)$, the *fuzzy Shapley value* $\phi(v)$ and the *fuzzy Weber set* $W(v)$ as follows (cf. [17]):

(i) $m^\sigma(v) = m^\sigma(cr(v))$ for each $\sigma \in \pi(N)$;
(ii) $\phi(v) = \frac{1}{|N|!} \sum_{\sigma \in \pi(N)} m^\sigma(v)$;
(iii) $W(v) = co\{m^\sigma(v) \mid \sigma \in \pi(N)\}$.

Note that $\phi(v) = \phi(cr(v))$, $W(v) = W(cr(v))$. Note further that for $i = \sigma(k)$, the i-th coordinate $m_i^\sigma(v)$ of the marginal vector $m^\sigma(v)$ is given by

$$m_i^\sigma(v) = v\left(\sum_{r=1}^{k} e^{\sigma(r)}\right) - v\left(\sum_{r=1}^{k-1} e^{\sigma(r)}\right).$$

One can identify each $\sigma \in \pi(N)$ with an n-step walk along the edges of the hypercube of fuzzy coalitions starting in $e^\emptyset$ and ending in e^N by passing the vertices $e^{\sigma(1)}, e^{\sigma(1)} + e^{\sigma(2)}, \ldots, \sum_{r=1}^{n-1} e^{\sigma(r)}$. The vector $m^\sigma(v)$ records the changes in value from vertex to vertex. The result in [124] that the core of a crisp game is a subset of the Weber set of the game can be extended for fuzzy games as we see in

Proposition 7.29. *Let $v \in FG^N$. Then $C(v) \subset W(v)$.*

Proof. By Remark 7.3 we have $C(v) \subset C(cr(v))$ and by Theorem 2.20, $C(cr(v)) \subset W(cr(v))$. Since $W(cr(v)) = W(v)$ we obtain $C(v) \subset W(v)$.

Note that the Weber set and the fuzzy Shapley value of a fuzzy game are very robust solution concepts since they are completely determined by the possibilities of crisp cooperation, regardless of what are the extra options that players could have as a result of graduating their participation rates. More specific solution concepts for fuzzy games will be introduced in the next sections of this chapter.

Inspired by [83] one can define the *diagonal value* $\delta(v)$ for a C^1-fuzzy game v (i.e. a game whose characteristic function is differentiable with continuous derivatives) as follows: for each $i \in N$ the i-th coordinate $\delta_i(v)$ of $\delta(v)$ is given by

$$\delta_i(v) = \int_0^1 D_i v(t, t, \ldots, t) dt,$$

where D_i is the partial derivative of v with respect to the i-th coordinate. According to Theorem 3.9 we have that for each crisp game $v \in G^N$:

$$\phi_i(v) = \delta_i(ml(v)) \text{ for each } i \in N.$$

The next example shows that for a fuzzy game v, $\delta(v)$ and $\phi(cr(v))$ may differ.

Example 7.30. Let $v \in FG^{\{1,2\}}$ with $v(s_1, s_2) = s_1(s_2)^2$ for each $s = (s_1, s_2) \in \mathcal{F}^{\{1,2\}}$. Then

$$m^{(1,2)} = (v(1,0) - v(0,0), v(1,1) - v(1,0)) = (0,1),$$

$$m^{(2,1)} = (v(1,1) - v(0,1), v(0,1) - v(0,0)) = (1,0);$$

so,

$$\phi(v) = \frac{1}{2}((0,1) + (1,0)) = \left(\frac{1}{2}, \frac{1}{2}\right).$$

Further,

$$D_1 v(s_1, s_2) = (s_2)^2, \; D_2 v(s_1, s_2) = 2 s_1 s_2;$$

so,

$$\delta_1(v) = \int_0^1 t^2 dt = \frac{1}{3}, \; \delta_2(v) = \int_0^1 2t^2 dt = \frac{2}{3}.$$

Hence,

$$\delta(v) = \left(\frac{1}{3}, \frac{2}{3}\right) \neq \left(\frac{1}{2}, \frac{1}{2}\right) = \phi(v).$$

The diagonal value is in fact the fuzzy value studied in [5], [6]. For extensions of this value the reader is referred to [31].

7.5 Path Solutions and the Path Solution Cover

Let us consider paths in the hypercube $[0,1]^N$ of fuzzy coalitions, which connect $e^{\emptyset}$ with e^N in a special way (cf. [19]).

Formally, a sequence $q = \langle p^0, p^1, \ldots, p^m \rangle$ of $m+1$ different points in $\mathcal{F}^N$ will be called a *path* (of length m) in $[0,1]^N$ if

(i) $p^0 = (0, 0, \ldots, 0)$, and $p^m = (1, 1, \ldots, 1)$;
(ii) $p^k \leq p^{k+1}$ for each $k \in \{0, \ldots, m-1\}$;
(iii) for each $k \in \{0, \ldots, m-1\}$, there is one player $i \in N$ (the acting player in point p^k) such that $\left(p^k\right)_j = \left(p^{k+1}\right)_j$ for all $j \in N \setminus \{i\}$, $\left(p^k\right)_i < \left(p^{k+1}\right)_i$.

For a path $q = \langle p^0, p^1, \ldots, p^m \rangle$ let us denote by $Q_i(q)$ the set of points p^k, where player i is acting, i.e. where $\left(p^k\right)_i < \left(p^{k+1}\right)_i$. Given a game $v \in FG^N$ and a path q, the payoff vector $x^q(v) \in \mathbb{R}^n$ corresponding to v and q has the i-th coordinate

$$x_i^q(v) = \sum_{k:p^k \in Q_i(q)} \left(v(p^{k+1}) - v(p^k) \right),$$

for each $i \in N$.

Given such a path $\langle p^0, p^1, \ldots, p^m \rangle$ of length m and $v \in FG^N$, one can imagine the situation, where the players in N, starting from non-cooperation $\left(p^0 = 0\right)$ arrive to full cooperation $\left(p^m = e^N\right)$ in m steps, where in each step one of the players increases his participation level. If the increase in value in such a step is given to the acting player, the resulting aggregate payoffs lead to the vector $x^q(v) = (x_i^q(v))_{i \in N}$. Note that $x^q(v)$ is an efficient vector, i.e. $\sum_{i=1}^n x_i^q(v) = v\left(e^N\right)$. We call $x^q(v)$ a *path solution*.

Let us denote by $Q(N)$ the set of paths in $[0,1]^N$. Then we denote by $Q(v)$ the convex hull of the set of path solutions and call it the *path solution cover*. Hence,

$$Q(v) = co \left\{ x^q(v) \in \mathbb{R}^n \mid q \in Q(N) \right\}.$$

Note that all paths $q \in Q(N)$ have length at least n. There are $n!$ paths with length exactly n; each of these paths corresponds to a situation where one by one the players – say in the order $\sigma(1), \ldots, \sigma(n)$ – increase their participation from level 0 to level 1. Let us denote such a path along n edges by q^σ. Then

$$q^\sigma = \left\langle 0, e^{\sigma(1)}, e^{\sigma(1)} + e^{\sigma(2)}, \ldots, e^N \right\rangle.$$

Clearly, $x^{q^\sigma}(v) = m^\sigma(v)$. Hence,

$$W(v) = co\left\{x^{q^\sigma}(v) \mid \sigma \text{ is an ordering of } N\right\} \subset Q(v).$$

According to Proposition 7.29, the core of a fuzzy game is a subset of the Weber set. Hence

Proposition 7.31. *For each $v \in FG^N$ we have $C(v) \subset W(v) \subset Q(v)$.*

Example 7.32. Let $v \in FG^{\{1,2\}}$ be given by $v(s_1, s_2) = s_1(s_2)^2 + s_1 + 2s_2$ for each $s = (s_1, s_2) \in \mathcal{F}^{\{1,2\}}$ and let $q \in Q(N)$ be the path of length 3 given by $\left\langle (0,0), \left(\frac{1}{3}, 0\right), \left(\frac{1}{3}, 1\right), (1,1)\right\rangle$. Then $x_1^q(v) = \left(v\left(\frac{1}{3}, 0\right) - v(0,0)\right) + \left(v(1,1) - v\left(\frac{1}{3}, 1\right)\right) = 1\frac{2}{3}$, $x_2^q(v) = v\left(\frac{1}{3}, 1\right) - v\left(\frac{1}{3}, 0\right) = 2\frac{1}{3}$. So $\left(1\frac{2}{3}, 2\frac{1}{3}\right) \in Q(v)$. The two shortest paths of length 2 given by $q^{(1,2)} = \langle (0,0), (1,0), (1,1)\rangle$ and $q^{(2,1)} = \langle (0,0), (0,1), (1,1)\rangle$ have payoff vectors $m^{(1,2)}(v) = (1,3)$, and $m^{(2,1)}(v) = (2,2)$, respectively.

Keeping in mind the interrelations among the Aubin core, the fuzzy Weber set and the path solution cover, one can try to introduce lower and upper bounds for payoff vectors in these sets. A lower (upper) bound is a payoff vector whose i-th coordinate is at most (at least) as good as the payoff given to player i when a "least desirable" ("most convenient") situation for him is achieved. By using pairs consisting of a lower bound and an upper bound, we obtain hypercubes which are catchers of the Aubin core, the fuzzy Weber set, and the path solution cover, respectively. In the next section we obtain compromise values for fuzzy games by taking a feasible compromise between the lower and upper bounds of the three catchers.

Formally, a *hypercube* in $\mathbb{R}^n$ is a set of vectors of the form

$$[a, b] = \{x \in \mathbb{R}^n \mid a_i \leq x_i \leq b_i \text{ for each } i \in N\},$$

where $a, b \in \mathbb{R}^n$, $a \leq b$ (and the order $\leq$ is the standard partial order in $\mathbb{R}^n$). The vectors a and b are called *bounding vectors* of the hypercube $[a, b]$, where, more explicitly, a is called the *lower vector* and b the *upper vector* of $[a, b]$. Given a set $A \subset \mathbb{R}^n$ we say that the hypercube $[a, b]$ is a *catcher* of A if $A \subset [a, b]$, and $[a, b]$ is called a *tight catcher* of A if there is no hypercube strictly included in $[a, b]$ which also catches A.

A hypercube of reasonable outcomes for a cooperative crisp game plays a role in [72] (see also [51]) and this hypercube can be seen as a tight catcher of the Weber set for crisp games (cf. Section 2.2). Also in [108] and [116] hypercubes are considered which are catchers of the core of crisp games.

Our aim is to introduce and study catchers of the Aubin core, the fuzzy Weber set and the path solution cover for games with a non-empty Aubin core, i.e. games which belong to FG_*^N.

Let us first introduce a core catcher

$$HC(v) = [l\left(C(v)\right), u\left(C(v)\right)]$$

for a game $v \in FG_*^N$, where for each $k \in N$:

$$l_k\left(C\left(v\right)\right) = \sup\left\{\varepsilon^{-1}v\left(\varepsilon e^k\right) \mid \varepsilon \in (0,1]\right\},$$

and

$$u_k\left(C\left(v\right)\right) = \inf\left\{\varepsilon^{-1}\left(v(e^N) - v\left(e^N - \varepsilon e^k\right)\right) \mid \varepsilon \in (0,1]\right\}.$$

Proposition 7.33. *For each $v \in FG_*^N$ and each $k \in N$:*

$$-\infty < l_k\left(C(v)\right) \leq u_k\left(C(v)\right) < \infty \text{ and } C(v) \subset HC(v).$$

Proof. Let $x \in C(v)$.

(i) For each $k \in N$ and $\varepsilon \in (0,1]$ we have

$$\begin{aligned} &v\left(e^N\right) - v\left(e^N - \varepsilon e^k\right) \\ &\geq \sum_{i \in N} x_i - \left((1-\varepsilon)\, x_k + \sum_{i \in N \setminus \{k\}} x_i\right) = \varepsilon x_k. \end{aligned}$$

So,

$$x_k \leq \varepsilon^{-1}\left(v(e^N) - v\left(e^N - \varepsilon e^k\right)\right)$$

implying that

$$x_k \leq u_k\left(C(v)\right) < \infty.$$

(ii) For each $\varepsilon \in (0,1]$ we have $\varepsilon x_k \geq v\left(\varepsilon e^k\right)$. Hence,

$$x_k \geq \sup\left\{\varepsilon^{-1}v\left(\varepsilon e^k\right) \mid \varepsilon \in (0,1]\right\} = l_k\left(C\left(v\right)\right) > -\infty.$$

By using (i) and (ii) one obtains the inequalities in the proposition and the fact that $HC(v)$ is a catcher of $C(v)$.

Now, we introduce for each $v \in FG_*^N$ a fuzzy variant $HW(v)$ of the hypercube of reasonable outcomes introduced in [72],

$$HW(v) = [l(W(v)), u(W(v))],$$

where for each $k \in N$:

$$l_k(W(v)) = \min\left\{v\left(e^{S\cup\{k\}}\right) - v\left(e^S\right) \mid S \subset N \setminus \{k\}\right\},$$

and

$$u_k(W(v)) = \max\left\{v\left(e^{S\cup\{k\}}\right) - v\left(e^S\right) \mid S \subset N \setminus \{k\}\right\}.$$

Then we have

Proposition 7.34. *For each $v \in FG_*^N$ the hypercube $HW(v)$ is a tight catcher of $W(v)$.*

Proof. Left to the reader.

Let us call a set $[a, b]$ with $a \leq b$, $a \in (\mathbb{R} \cup \{-\infty\})^n$ and $b \in (\mathbb{R} \cup \{\infty\})^n$ a *generalized* hypercube.

Now, we introduce for $v \in FG_*^N$ the generalized hypercube

$$HQ(v) = [l(Q(v)), u(Q(v))],$$

which catches the path solution cover $Q(v)$ as we see in Theorem 7.35, where for $k \in N$:

$$l_k(Q(v)) = \\ \inf\left\{\varepsilon^{-1}\left(v\left(s + \varepsilon e^k\right) - v(s)\right) \mid s \in \mathcal{F}^N, s_k < 1, \varepsilon \in (0, 1 - s_k]\right\},$$

$$u_k(Q(v)) = \\ \sup\left\{\varepsilon^{-1}\left(v\left(s + \varepsilon e^k\right) - v(s)\right) \mid s \in \mathcal{F}^N, s_k < 1, \varepsilon \in (0, 1 - s_k]\right\},$$

where $l_k(Q(v)) \in [-\infty, \infty)$ and $u_k(Q(v)) \in (-\infty, \infty]$.

Note that $u(Q(v)) \geq u(C(v))$, $l(Q(v)) \leq l(C(v))$.

Theorem 7.35. *For $v \in FG_*^N$, $HQ(v)$ is a catcher of $Q(v)$.*

Proof. This assertion follows from the fact that for each path $q \in Q(N)$ and any $i \in N$

$$x_i^q(v) = \sum_{k:p^k \in Q_i(q)} \left(v\left(p^k + \left(p_i^{k+1} - p_i^k\right) e^i\right) - v\left(p^k\right)\right)$$
$$\leq \sum_{k:p^k \in Q_i(q)} \left(p_i^{k+1} - p_i^k\right) u_i\left(Q(v)\right) = u_i\left(Q(v)\right),$$

and, similarly,

$$x_i^q(v) \geq l_i\left(Q(v)\right).$$

Note that the lower and upper bounds of the catcher of the fuzzy Weber set are obtained by using a finite number of value differences, where only coalitions corresponding to crisp coalitions play a role. The calculation of the lower and upper bounds of the catchers of the Aubin core and of the path solution cover is based on infinite value differences.

7.6 Compromise Values

We introduce now for fuzzy games compromise values of σ-type and of τ-type with respect to each of the set-valued solutions C, W and Q. For the first type use is made directly of the bounding vectors of $HC(v), HW(v)$ and $HQ(v)$, while for the τ-type compromise values the upper vector is used together with the remainder vector derived from the upper vector (cf. [19]).

To start with the first type, consider a hypercube $[a, b]$ in $\mathbb{R}^n$ and a game $v \in FG_*^N$ such that the hypercube contains at least one efficient vector, i.e.

$$[a, b] \cap \left\{ x \in \mathbb{R}^n \mid \sum_{i=1}^{n} x_i = v\left(e^N\right)\right\} \neq \emptyset.$$

Then there is a unique point $c(a, b)$ on the line through a and b which is also efficient in the sense that $\sum_{i=1}^n c_i(a, b) = v\left(e^N\right)$. So $c(a, b)$ is the convex combination of a and b, which is efficient. We call $c(a, b)$ the *feasible compromise* between a and b.

Now we introduce the following three σ-like compromises for $v \in FG_*^N$:

$$val_C^\sigma(v) = c\left(HC(v)\right) = c\left(\left[l\left(C(v)\right), u\left(C(v)\right)\right]\right),$$
$$val_W^\sigma(v) = c\left(HW(v)\right) = c\left(\left[l\left(W(v)\right), u\left(W(v)\right)\right]\right),$$

and

$$val_Q^\sigma(v) = c\left(HQ(v)\right) = c\left(\left[l\left(Q(v)\right), u\left(Q(v)\right)\right]\right)$$

if the generalized hypercube $HQ(v)$ is a hypercube.

Note that

$$\emptyset \neq C(v) \subset HC(v) \subset HQ(v), \tag{7.1}$$

and

$$\emptyset \neq C(v) \subset W(v) \subset HW(v), \tag{7.2}$$

so all hypercubes contain efficient vectors and the first two compromise value vectors are always well defined.

For the τ-like compromise values inspired by the definition of minimal right vectors for crisp games (cf. [11], [44], and Section 2.2) we define the fuzzy minimal right operator $m^v : \mathbb{R}^n \to \mathbb{R}^n$ for $v \in FG_*^N$ by

$$m_i^v(z) = \sup\left\{ s_i^{-1}\left(v(s) - \sum_{j\in N\setminus\{i\}} s_j z_j\right) \mid s \in \mathcal{F}_0^N, s_i > 0\right\}$$

for each $i \in N$ and each $z \in \mathbb{R}^n$.

The following proposition shows that m^v assigns to each upper bound z of the Aubin core (i.e. $z \geq x$ for each $x \in C(v)$) a lower bound $m^v(z)$ of $C(v)$, called the *remainder vector* corresponding to z.

Proposition 7.36. *Let $v \in FG_*^N$ and let $z \in \mathbb{R}^n$ be an upper bound of $C(v)$. Then $m^v(z)$ is a lower bound of $C(v)$.*

Proof. Let $i \in N$ and $x \in C(v)$. For each $s \in \mathcal{F}^N$ with $s_i > 0$ we have

$$\begin{aligned} s_i^{-1}\left(v(s) - \sum_{j\in N\setminus\{i\}} s_j z_j\right) &\leq s_i^{-1}\left(\sum_{j\in N} s_j x_j - \sum_{j\in N\setminus\{i\}} s_j z_j\right) \\ &= x_i + s_i^{-1} \sum_{j\in N\setminus\{i\}} s_j\left(x_j - z_j\right) \\ &\leq x_i, \end{aligned}$$

where the first inequality follows from $x \in C(v)$ and the second inequality from the fact that z is an upper bound for $C(v)$, and then $z \geq x$. Hence $m_i^v(z) \leq x_i$ for each $i \in N$, which means that $m^v(z)$ is a lower bound for $C(v)$.

Now we are able to introduce the τ-like compromise values taking into account that all upper vectors of $HC(v), HW(v)$ and $HQ(v)$ are upper bounds for the Aubin core of $v \in FG_*^N$ as follows from (7.1) and (7.2).

So the following definitions make sense for $v \in FG_*^N$:

$$val_C^{\mathcal{T}}(v) = c\left(\left[m^v\left(u\left(C(v)\right)\right), u\left(C(v)\right)\right]\right),$$

$$val_W^{\mathcal{T}}(v) = c\left(\left[m^v\left(u\left(W(v)\right)\right), u\left(W(v)\right)\right]\right),$$

and

$$val_Q^{\mathcal{T}}(v) = c\left(\left[m^v\left(u\left(Q(v)\right)\right), u\left(Q(v)\right)\right]\right)$$

if the generalized hypercube $HQ(v)$ is a hypercube.

The compromise value $val_C^{\mathcal{T}}(v)$ is in the spirit of the τ-value introduced in [108] for cooperative crisp games (see Section 3.2), and the compromise value $val_W^{\mathcal{T}}(v)$ is in the spirit of the χ-value in [12], the μ-value in [57] and one of the values in [29] and [30] for cooperative crisp games.

8

Convex Fuzzy Games

An interesting class of fuzzy games is generated when the notion of convexity is considered. Convex fuzzy games can be successfully used for solving sharing problems arising from many economic situations where "cooperation" is the main benefit/cost savings generator; all the solution concepts treated in Chapter 7 have nice properties for such games. Moreover, for convex fuzzy games one can use additional sharing rules which are based on more specific solution concepts like participation monotonic allocation schemes and egalitarian solutions.

8.1 Basic Characterizations

Let $v : [0,1]^n \to \mathbb{R}$. Then v satisfies

(i) *supermodularity* (SM) if

$$v(s \vee t) + v(s \wedge t) \geq v(s) + v(t) \text{ for all } s, t \in [0,1]^N, \text{and} \quad (8.1)$$

(ii) *coordinate-wise convexity* (CwC) if for each $i \in N$ and each $s^{-i} \in [0,1]^{N \setminus \{i\}}$ the function $g_{s^{-i}} : [0,1] \to \mathbb{R}$ with $g_{s^{-i}}(t) = v(s^{-i} \parallel t)$ for each $t \in [0,1]$ is a convex function (see page 78).

Now, we introduce our definition of convex cooperative fuzzy games (cf. [17]).

Definition 8.1. *Let $v \in FG^N$. Then v is called a* ***convex fuzzy game*** *if $v : [0,1]^N \to \mathbb{R}$ satisfies SM and CwC.*

Remark 8.2. Convex fuzzy games form a convex cone.

Remark 8.3. For a weaker definition of a convex fuzzy game we refer to [120], where only the supermodularity property is used.

As shown in Proposition 8.4, a nice example of convex fuzzy games is a unanimity game in which the minimal wining coalition corresponds to a crisp-like coalition.

Proposition 8.4. *Let* $u_t \in FG^N$ *be the unanimity game based on the fuzzy coalition* $t \in \mathcal{F}_0^N$. *Then the game* u_t *is convex if and only if* $t = e^T$ *for some* $T \in 2^N \setminus \{\emptyset\}$.

Proof. Suppose $t \neq e^T$ for some $T \in 2^N \setminus \{\emptyset\}$. Then there is a $k \in N$ such that $\varepsilon = \min\{t_k, 1 - t_k\} > 0$ and $0 = u_t(t + \varepsilon e^k) - u_t(t) < u_t(t) - u_t(t - \varepsilon e^k) = 1$, implying that u_t is not convex.

Conversely, suppose that $t = e^T$ for some $T \in 2^N \setminus \{\emptyset\}$. Then we show that u_t has the supermodularity property and the coordinate-wise convexity property. Take $s, k \in \mathcal{F}^N$. We can distinguish three cases.

(1) If $u_t(s \vee k) + u_t(s \wedge k) = 2$, then $u_t(s \wedge k) = 1$. Thus, the supermodularity condition (8.1) follows from $u_t(s) + u_t(k) \geq 2u_t(s \wedge k) = 2$.
(2) If $u_t(s \vee k) + u_t(s \wedge k) = 0$, then $u_t(s \vee k) = 0$. Hence $u_t(s) + u_t(k) \leq 2u_t(s \vee k) = 0$, $u_t(s) + u_t(k) = 0$.
(3) If $u_t(s \vee k) + u_t(s \wedge k) = 1$, then $u_t(s \vee k) = 1$ and $u_t(s \wedge k) = 0$. Therefore, $u_t(s)$ or $u_t(k)$ must be equal to 0, and, consequently, $u_t(s) + u_t(k) \leq 1$ and the supermodularity condition (8.1) is fulfilled.

Hence, the supermodularity property holds for u_{e^T}.

To prove the coordinate-wise convexity of u_{e^T}, note that all functions $g_{s_{-i}}$ in the definition of coordinate-wise convexity are convex because they are either constant with value 0 or with value 1, or they have value 0 on $[0, 1)$ and value 1 in 1. So, u_{e^T} is a convex game.

In the following the set of convex fuzzy games with player set N will be denoted by CFG^N. Clearly, $CFG^N \subset FG^N$. Remember that the set of convex crisp games with player set N was denoted by CG^N.

Proposition 8.5. *Let* $v \in CFG^N$. *Then* $cr(v) \in CG^N$.

Proof. We will prove that $cr(v)$ satisfies SM for crisp games (cf. (5.1)). Take $S, T \in 2^N$ and apply SM for fuzzy games (8.1) with $e^S, e^T, e^{S \cup T}, e^{S \cap T}$ in the roles of $s, t, s \vee t, s \wedge t$, respectively, obtaining

$$cr(v)\,(S \cup T) + cr(v)(S \cap T) \geq cr(v)(S) + cr(v)(T).$$

The next property for convex fuzzy games is related with the increasing marginal contribution property for players in crisp games (cf. Theorem 5.10(iii)). It states that a level increase of a player in a fuzzy coalition has more beneficial effect in a larger coalition than in a smaller coalition.

Proposition 8.6. *Let $v \in CFG^N$. Let $i \in N$, $s^1, s^2 \in \mathcal{F}^N$ with $s^1 \leq s^2$ and let $\varepsilon \in \mathbb{R}_+$ with $0 \leq \varepsilon \leq 1 - s_i^2$. Then*

$$v\left(s^1 + \varepsilon e^i\right) - v\left(s^1\right) \leq v\left(s^2 + \varepsilon e^i\right) - v\left(s^2\right). \tag{8.2}$$

Proof. Suppose $N = \{1, \ldots, n\}$. Define the fuzzy coalitions $c^0, c^1, \ldots, c^n$ by $c^0 = s^1$, and $c^k = c^{k-1} + \left(s_k^2 - s_k^1\right) e^k$ for $k \in \{1, \ldots, n\}$. Then $c^n = s^2$. To prove (8.2) it is sufficient to show that for each $k \in \{1, \ldots, n\}$ the inequality $\left(I^k\right)$ holds

$$v\left(c^k + \varepsilon e^i\right) - v\left(c^k\right) \geq v\left(c^{k-1} + \varepsilon e^i\right) - v\left(c^{k-1}\right). \tag{I^k}$$

Note that $\left(I^i\right)$ follows from the coordinate-wise convexity of v and $\left(I^k\right)$ for $k \neq i$ follows from SM with $c^{k-1} + \varepsilon e^i$ in the role of s and c^k in the role of t. Then $s \vee t = c^k + \varepsilon e^i$, $s \wedge t = c^{k-1}$.

Also an analogue of the increasing marginal contribution property for coalitions (cf. Theorem 5.10(ii)) holds as we see in

Proposition 8.7. *Let $v \in CFG^N$. Let $s, t \in \mathcal{F}^N$ and $z \in \mathbb{R}_+^n$ such that $s \leq t \leq t + z \leq e^N$. Then*

$$v(s + z) - v(s) \leq v(t + z) - v(t). \tag{8.3}$$

Proof. For each $k \in \{1, \ldots, n\}$ it follows from Proposition 8.6 (with $s + \sum_{r=1}^{k-1} z_r e^r$ in the role of s^1, $t + \sum_{r=1}^{k-1} z_r e^r$ in the role of s^2, k in the role of i, and z_k in the role of ε) that

$$\begin{aligned} & v\left(s + \sum_{r=1}^{k} z_r e^r\right) - v\left(s + \sum_{r=1}^{k-1} z_r e^r\right) \\ & \leq v\left(t + \sum_{r=1}^{k} z_r e^r\right) - v\left(t + \sum_{r=1}^{k-1} z_r e^r\right). \end{aligned}$$

By adding these n inequalities we obtain inequality (8.3).

The next proposition introduces a characterizing property for convex fuzzy games which we call the *increasing average marginal return property* (IAMR). This property expresses the fact that for a convex game an increase in participation level of any player in a smaller coalition yields per unit of level less than an increase in a larger coalition under the condition that the reached level of participation in the first case is still not bigger than the reached participation level in the second case.

Proposition 8.8. *Let $v \in CFG^N$. Let $i \in N$, $s^1, s^2 \in \mathcal{F}^N$ with $s^1 \le s^2$ and let $\varepsilon_1, \varepsilon_2 > 0$ with $s_i^1 + \varepsilon_1 \le s_i^2 + \varepsilon_2 \le 1$. Then*

$$\varepsilon_1^{-1}\left(v\left(s^1 + \varepsilon_1 e^i\right) - v\left(s^1\right)\right) \le \varepsilon_2^{-1}\left(v\left(s^2 + \varepsilon_2 e^i\right) - v\left(s^2\right)\right). \quad (8.4)$$

Proof. From Proposition 8.6 (with s^1, $\left(s^2 + \left(s_i^1 - s_i^2\right)e^i\right)$ and ε_1 in the roles of s^1, s^2 and ε, respectively) it follows that

$$\begin{aligned}\varepsilon_1^{-1}\left(v\left(s^2 + \left(s_i^1 - s_i^2 + \varepsilon_1\right)e^i\right) - v\left(s^2 + \left(s_i^1 - s_i^2\right)e^i\right)\right)\\ \ge \varepsilon_1^{-1}\left(v\left(s^1 + \varepsilon_1 e^i\right) - v\left(s^1\right)\right).\end{aligned}$$

Further, from CwC (by noting that $s_i^2 + \varepsilon_2 \ge s_i^2 + \left(s_i^1 - s_i^2 + \varepsilon_1\right)$, $s_i^2 \ge s_i^2 + \left(s_i^1 - s_i^2\right)$) it follows that

$$\begin{aligned}&\varepsilon_2^{-1}\left(v\left(s^2 + \varepsilon_2 e^i\right) - v\left(s^2\right)\right)\\ &\ge \varepsilon_1^{-1}\left(v\left(s^2 + \left(s_i^1 - s_i^2 + \varepsilon_1\right)e^i\right) - v\left(s^2 + \left(s_i^1 - s_i^2\right)e^i\right)\right),\end{aligned}$$

resulting in the desired inequality.

Theorem 8.9. *Let $v \in FG^N$. Then the following assertions are equivalent:*

(i) $v \in CFG^N$;
(ii) v satisfies IAMR.

Proof. We know from Proposition 8.8 that a convex game satisfies IAMR. On the other hand it is clear that IAMR implies the CwC. Hence, we only have to prove that IAMR implies SM. So, given $s, t \in \mathcal{F}^N$ we have to prove inequality (8.1).

Let $P = \{i \in N \mid t_i < s_i\}$. If $P = \emptyset$, then (8.1) follows from the fact that $s \vee t = t, s \wedge t = s$. For $P \ne \emptyset$, arrange the elements of P in a sequence $\sigma(1), \ldots, \sigma(p)$, where $p = |P|$, and put $\varepsilon_{\sigma(k)} = s_{\sigma(k)} - t_{\sigma(k)} > 0$ for $k \in \{1, \ldots, p\}$. Note that in this case

$$s = s \wedge t + \sum_{k=1}^{p} \varepsilon_{\sigma(k)} e^{\sigma(k)}, \quad s \vee t = t + \sum_{k=1}^{p} \varepsilon_{\sigma(k)} e^{\sigma(k)}.$$

Hence,

$$\begin{aligned}&v(s) - v(s \wedge t) =\\ &\textstyle\sum_{r=1}^{p}\left(v\left(s \wedge t + \sum_{k=1}^{r} \varepsilon_{\sigma(k)} e^{\sigma(k)}\right) - v\left(s \wedge t + \sum_{k=1}^{r-1} \varepsilon_{\sigma(k)} e^{\sigma(k)}\right)\right),\end{aligned}$$

and

$$\begin{aligned}&v(s \vee t) - v(t) =\\ &\textstyle\sum_{r=1}^{p}\left(v\left(t + \sum_{k=1}^{r} \varepsilon_{\sigma(k)} e^{\sigma(k)}\right) - v\left(t + \sum_{k=1}^{r-1} \varepsilon_{\sigma(k)} e^{\sigma(k)}\right)\right).\end{aligned}$$

From these equalities the relation (8.1) follows because IAMR implies for each $r \in \{1, \ldots, p\}$:

$$v\left(s \wedge t + \sum_{k=1}^{r} \varepsilon_{\sigma(k)} e^{\sigma(k)}\right) - v\left(s \wedge t + \sum_{k=1}^{r-1} \varepsilon_{\sigma(k)} e^{\sigma(k)}\right)$$
$$\leq v\left(t + \sum_{k=1}^{r} \varepsilon_{\sigma(k)} e^{\sigma(k)}\right) - v\left(t + \sum_{k=1}^{r-1} \varepsilon_{\sigma(k)} e^{\sigma(k)}\right).$$

Next, we study the implications of two other properties a function $v : [0,1]^N \to \mathbb{R}$ may satisfy (cf. [113]). The first one we call *monotonicity of the first partial derivatives property* (MOPAD), and the second one is the *directional convexity property* (DICOV) introduced in [70].

Let $i \in N$ and $s \in [0,1]^N$. We say that the left derivative $D_i^- v(s)$ of v in the i-th direction at s with $0 < s_i \leq 1$ exists if $\lim_{\substack{\varepsilon \to 0 \\ \varepsilon > 0}} \varepsilon^{-1}(v(s) - v(s - \varepsilon e^i))$ exists and is finite; then

$$D_i^- v(s) := \lim_{\substack{\varepsilon \to 0 \\ \varepsilon > 0}} \varepsilon^{-1}(v(s) - v(s - \varepsilon e^i)).$$

Similarly, the right derivative of v in the i-th direction at s with $0 \leq s_i < 1$, denoted by $D_i^+ v(s)$, is $\lim_{\substack{\varepsilon \to 0 \\ \varepsilon > 0}} \varepsilon^{-1}(v(s + \varepsilon e^i) - v(s))$ if this limit exists and is finite. We put for convenience $D_i^- v(s) = -\infty$ if $s_i = 0$ and $D_i^+ v(s) = \infty$ if $s_i = 1$.

Definition 8.10. *We say that $v : [0,1]^N \to \mathbb{R}$ satisfies* ***MOPAD*** *if for each $i \in N$ the following four conditions hold:*

(Ma) $D_i^- v(s)$ exists for each $s \in [0,1]^N$ with $0 < s_i \leq 1$;
(Mb) $D_i^+ v(s)$ exists for each $s \in [0,1]^N$ with $0 \leq s_i < 1$;
(Mc) $D_i^- v(s) \leq D_i^+ v(s)$;
(Md) $D_i^- v(s^1) \leq D_i^- v(s^2)$ and $D_i^+ v(s^1) \leq D_i^+ v(s^2)$ for each $s^1, s^2 \in [0,1]^N$ with $s^1 \leq s^2$.

Definition 8.11. *Let $[a,b] = \{x \in [0,1]^N \mid a_i \leq x_i \leq b_i$ for each $i \in N\}$. We say that $v : [0,1]^N \to \mathbb{R}$ satisfies* ***DICOV*** *if for each $a, b \in [0,1]^N$ with $a \leq b$ and each pair $c, d \in [a,b]$ with $c + d = a + b$ it follows that*

$$v(a) + v(b) \geq v(c) + v(d).$$

Lemma 8.12. *Let $s \in [0,1]^N$ and $i \in N$ with $0 \leq s_i < 1$ and $0 < \varepsilon_1 \leq \varepsilon_2 < \varepsilon_3 \leq 1 - s_i$. If $v : [0,1]^N \to \mathbb{R}$ satisfies MOPAD then*

$$\varepsilon_1^{-1}(v(s + \varepsilon_1 e^i) - v(s)) \leq (\varepsilon_3 - \varepsilon_2)^{-1}(v(s + \varepsilon_3 e^i) - v(s + \varepsilon_2 e^i)).$$

Proof. From the definition of the left derivative and by (M4) it follows
$\varepsilon_1^{-1}(v(s+\varepsilon_1 e^i)-v(s)) = \varepsilon_1^{-1}\int_0^{\varepsilon_1} D_i^- v(s+xe^i)dx$
$\leq \varepsilon_1^{-1}\int_0^{\varepsilon_1} D_i^- v(s+\varepsilon_1 e^i)dx = D_i^- v(s+\varepsilon_1 e^i) \leq D_i^- v(s+\varepsilon_2 e^i)$
$= (\varepsilon_3-\varepsilon_2)^{-1}\int_{\varepsilon_2}^{\varepsilon_3} D_i^- v(s+\varepsilon_2 e^i)dx \leq (\varepsilon_3-\varepsilon_2)^{-1}\int_{\varepsilon_2}^{\varepsilon_3} D_i^- v(s+xe^i)dx$
$=(\varepsilon_3-\varepsilon_2)^{-1}(v(s+\varepsilon_3 e^i)-v(s+\varepsilon_2 e^i))$.

The following theorem establishes the equivalence among the introduced properties provided that the characteristic function v is continuous.

Theorem 8.13. *Let $v : [0,1]^N \to \mathbb{R}$ be a continuous function. The following assertions are equivalent:*

(i) v satisfies SM and CwC;
(ii) v satisfies MOPAD;
(iii) v satisfies IAMR;
(iv) v satisfies DICOV.

Proof. (i) $\to$ (ii): The validity of (M1), (M2), and (M3) in the definition of MOPAD follows by CwC. To prove (M4) note first that for s^1 with $s_i^1 = 0$ we have $D_i^- v(s^1) = -\infty \leq D_i^- v(s^2)$. If $s_i^1 > 0$, then CwC implies (cf. Proposition 8.6) that

$$v(s^1) - v(s^1 - \varepsilon e_i) \leq v(s^2) - v(s^2 - \varepsilon e_i)$$

for $\varepsilon > 0$ such that $s_i^1 - \varepsilon \geq 0$. By multiplying the left and right sides of the above inequality with ε^{-1} and then taking the limit for ε going to 0, we obtain $D_i^- v(s^1) \leq D_i^- v(s^2)$. The second inequality in (M4) can be proved in a similar way.

(ii) $\to$ (iii): Suppose that v satisfies MOPAD. We have to prove that for each $a, b \in [0,1]^N$ with $a \leq b$, each $i \in N$, $\delta \in (0, 1-a_i]$, $\varepsilon \in (0, 1-b_i]$ such that $a_i + \delta \leq b_i + \varepsilon \leq 1$ it follows that

$$\delta^{-1}(v(a+\delta e^i) - v(a)) \leq \varepsilon^{-1}(v(b+\varepsilon e^i) - v(b)). \tag{8.5}$$

Take $a, b, i, \delta, \varepsilon$ as above. Let $c = a + (b_i - a_i)e^i$ and $d = b + (a_i + \delta - b_i)e^i$. We consider two cases:

(α) $a_i + \delta \leq b_i$. Then
$\delta^{-1}(v(a+\delta e^i)-v(a)) = \delta^{-1}\int_0^{\delta} D_i^- v(a+xe^i)dx \leq$
$\delta^{-1}\int_0^{\delta} D_i^- v(a+\delta e^i)dx = D_i^- v(a+\delta e^i) \leq D_i^- v(b) \leq$
$\varepsilon^{-1}\int_0^{\varepsilon} D_i^- v(b+xe^i)dx = \varepsilon^{-1}(v(b+\varepsilon e^i)-v(b))$,
where the inequalities follow by (M4).

(β) $a_i + \delta \in (b_i, b_i + \varepsilon]$. Then

$\delta^{-1}(v(a+\delta e^i)-v(a)) = \delta^{-1}(v(c)-v(a)) + \delta^{-1}(v(a+\delta e^i)-v(c)) \leq$
$\delta^{-1}(c_i-a_i)(a_i+\delta-c_i)^{-1}(v(a+\delta e^i)-v(c)) + \delta^{-1}(v(a+\delta e^i)-v(c)) =$
$(a_i+\delta-c_i)^{-1}(v(a+\delta e^i)-v(c))$,
where the inequality follows by Lemma 8.12, with $s = a$, $\varepsilon_1 = \varepsilon_2 = c_i - a_i < \delta = \varepsilon_3$.

Thus, we have

$$\delta^{-1}(v(a+\delta e^i)-v(a)) \leq (a_i+\delta-c_i)^{-1}(v(a+\delta e^i)-v(c)). \quad (8.6)$$

Similarly, by applying Lemma 8.12 with $s = b$, $\varepsilon_1 = \varepsilon_2 = d_i - b_i < \varepsilon = \varepsilon_3$, it follows that

$$(d_i-b_i)^{-1}(v(d)-v(b)) \leq \varepsilon^{-1}(v(b+\varepsilon e^i)-v(b)). \quad (8.7)$$

Further, using (M4) with $(a_{-i},t) = (a_1,\ldots,a_{i-1},t,a_{i+1},\ldots,a_n)$ and $(b_{-i},t) = (b_1,\ldots,b_{i-1},t,b_{i+1},\ldots,b_n)$ in the role of s^1 and s^2, respectively, and the equalities $a_i+\delta = d_i$ and $c_i = b_i$ we obtain
$v\left(a+\delta e^i\right) - v(c) = \int_{c_i}^{a_i+\delta} D_i^- v(a_{-i},t)dt \leq \int_{b_i}^{d_i} D_i^- v(b_{-i},t)dt \leq v(d)-v(b)$.

Now, since $a_i+\delta-c_i = d_i-b_i$, we have

$$(a_i+\delta-c_i)^{-1}\left(v\left(a+\delta e^i\right)-v(c)\right) \leq (d_i-b_i)^{-1}\left(v(d)-v(b)\right). \quad (8.8)$$

Combining (8.6) and (8.7) via (8.8) (by using the transitivity property of the inequality relation), (8.5) follows.

(iii) $\to$ (iv): Assume that v satisfies IAMR. Take $a, b \in [0,1]^N$ with $a \leq b$ and $c, d \in [a,b]$ with $c+d = a+b$. Define $h = b-c$. Then $b = c+h$ and $d = a+h$. We have
$v(b)-v(c) = \sum_{r=1}^n (v(c+\sum_{i=1}^r h_i e^i) - v(c+\sum_{i=1}^{r-1} h_i e^i)) \geq$
$\sum_{r=1}^n (v(a+\sum_{i=1}^r h_i e^i) - v(a+\sum_{i=1}^{r-1} h_i e^i)) = f(d)-f(a)$,
where the inequality follows by applying IAMR n times.

(iv) $\to$ (i): Let v satisfy DICOV. To prove that v satisfies CwC as well, note that for $s^{-i} \in [0,1]^{N\setminus\{i\}}$ and $0 \leq p < \frac{1}{2}(p+q) < q \leq 1$, we have $(s^{-i} \parallel \frac{1}{2}(p+q)) \in [(s^{-i} \parallel p), (s^{-i} \parallel q)]$. So, (iv) with $a = (s^{-i} \parallel p)$, $b = (s^{-i} \parallel q)$, $c = d = (s^{-i} \parallel \frac{1}{2}(p+q))$ implies $v(s^{-i} \parallel p) + v(s^{-i} \parallel q) \geq 2v(s^{-i} \parallel \frac{1}{2}(p+q))$.

To prove that v satisfies SM, let $c, d \in [0,1]^N$. Then $c, d \in [c \wedge d, c \vee d]$. By (iv) one obtains $v(c \vee d) + v(c \wedge d) \geq v(c) + v(d)$.

Finally, we introduce a fifth property that allows for a very simple characterization of a convex fuzzy game. It requires that all second partial derivatives of $v : [0,1]^N \to \mathbb{R}$ are non-negative (NNSPAD).

Definition 8.14. *Let* $v \in C^2$. *Then* v *satisfies* ***NNSPAD*** *on* $[0,1]^N$ *if for all* $i, j \in N$ *we have*

$$\frac{\partial^2 v}{\partial s_i \partial s_j} \geq 0.$$

Obviously, the properties MOPAD and NNSPAD are equivalent on the class of C^2-functions.

Remark 8.15. In [101] it is shown that if $v \in C^2$, then DICOV implies NNSPAD.

Remark 8.16. For each $r \in \{1, \ldots, m\}$, let μ_r be defined for each $s \in [0,1]^N$ by $\mu_r(s) = \sum_{i \in N} s_i \mu_r(i\})$ with $\mu_r(i\}) \geq 0$ for each $i \in N$ and $\sum_{i \in N} \mu_r(i\}) \leq 1$. If $f \in C^1$ satisfies DICOV, then v with $v(s) = f(\mu_1(s), \ldots, \mu_m(s))$ is a convex game (cf. [70]).

8.2 Egalitarianism in Convex Fuzzy Games

In this section we are interested in introducing an egalitarian solution for convex fuzzy games. We do this in a constructive way by adjusting the classical Dutta-Ray algorithm for a convex crisp game (cf. [46]).

As mentioned in Subsection 5.2.3, at each step of the Dutta-Ray algorithm for convex crisp games a largest element exists. Note that for the crisp case the supermodularity of the characteristic function is equivalent to the convexity of the corresponding game.

Although the cores of a convex fuzzy game and its related (convex) crisp game coincide and the Dutta-Ray constrained egalitarian solution is a core element, finding the egalitarian solution of a convex fuzzy game is a task on itself. As we show in Lemma 8.17, supermodularity of a fuzzy game implies a semilattice structure of the corresponding (possibly infinite) set of fuzzy coalitions with maximal average worth (cf. (6.1)), but it is not enough to ensure the existence of a maximal element. Different difficulties which can arise in fuzzy games satisfying only the supermodularity property are illustrated by means of three examples. According to Lemma 8.21 it turns out that adding coordinate-wise convexity to supermodularity is sufficient for the existence of such a maximal element; moreover, this element corresponds to a crisp coalition. Then, a simple method becomes available to calculate the egalitarian solution of a convex fuzzy game (cf. [18]).

Lemma 8.17. *Let* $v \in FG^N$ *be a supermodular game. Then the set*

$$A(N,v) := \left\{ t \in \mathcal{F}_0^N \mid \alpha(t,v) = \sup_{s \in \mathcal{F}_0^N} \alpha(s,v) \right\}$$

is closed with respect to the join operation $\vee$.

Proof. Let $\overline{\alpha} = \sup_{s\in\mathcal{F}_0^N} \alpha(s,v)$. If $\overline{\alpha} = \infty$, then $A(N,v) = \emptyset$, so $A(N,v)$ is closed w.r.t. the join operation.

Suppose now $\overline{\alpha} \in \mathbb{R}$. Take $t^1, t^2 \in A(N,v)$. We have to prove that $t^1 \vee t^2 \in A(N,v)$, that is $\alpha\left(t^1 \vee t^2, v\right) = \overline{\alpha}$.

Since $v\left(t^1\right) = \overline{\alpha}\left\lceil t^1 \right\rfloor$ and $v\left(t^2\right) = \overline{\alpha}\left\lceil t^2 \right\rfloor$ we obtain

$$\begin{aligned}\overline{\alpha}\left\lceil t^1\right\rfloor + \overline{\alpha}\left\lceil t^2\right\rfloor &= v\left(t^1\right) + v\left(t^2\right) \le v\left(t^1 \vee t^2\right) + v\left(t^1 \wedge t^2\right)\\ &\le \overline{\alpha}\left\lceil t^1 \vee t^2\right\rfloor + \overline{\alpha}\left\lceil t^1 \wedge t^2\right\rfloor = \overline{\alpha}\left\lceil t^1\right\rfloor + \overline{\alpha}\left\lceil t^2\right\rfloor,\end{aligned}$$

where the first inequality follows from SM and the second inequality follows from the definition of $\overline{\alpha}$ and the fact that $v(e^{\emptyset}) = 0$. This implies that $v\left(t^1 \vee t^2\right) = \overline{\alpha}\left\lceil t^1 \vee t^2\right\rfloor$, so $t^1 \vee t^2 \in A(N,v)$.

We can conclude from the proof of Lemma 8.17 that in case $t^1, t^2 \in A(N,v)$ not only $t^1 \vee t^2 \in A(N,v)$ but also $t^1 \wedge t^2 \in A(N,v)$ if $t^1 \wedge t^2 \neq e^{\emptyset}$. Further, $A(N,v)$ is closed w.r.t. finite "unions", where $t^1 \vee t^2$ is seen as the "union" of t^1 and t^2.

If we try to introduce in a way similar to that of [46] an egalitarian rule for supermodular fuzzy games, then problems may arise since the set of non-empty fuzzy coalitions is infinite and it is not clear if there exists a maximal fuzzy coalition with "maximum value per unit of participation level". To be more precise, if $v \in FG^N$ is a supermodular fuzzy game then crucial questions are:

(1) Is $\sup_{s\in\mathcal{F}_0^N} \alpha(s,v)$ finite or not? Example 8.18 presents a fuzzy game for which $\sup_{s\in\mathcal{F}_0^N} \alpha(s,v)$ is infinite.

(2) In case that $\sup_{s\in\mathcal{F}_0^N} \alpha(s,v)$ is finite, is there a $t \in \mathcal{F}_0^N$ s.t. $\alpha(t,v) = \sup_{s\in\mathcal{F}_0^N} \alpha(s,v)$? A fuzzy game for which the set $\arg\sup_{s\in\mathcal{F}_0^N} \alpha(s,v)$ is empty is given in Example 8.19. Note that if the set $\arg\sup_{s\in\mathcal{F}_0^N} \alpha(s,v)$ is non-empty then $\sup_{s\in\mathcal{F}_0^N} \alpha(s,v) = \max_{s\in\mathcal{F}_0^N} \alpha(s,v)$.

(3) Let $\geq$ be the standard partial order on $[0,1]^N$. Suppose that $\max_{s\in\mathcal{F}_0^N} \alpha(s,v)$ exists. Does the set $\arg\max_{s\in\mathcal{F}_0^N} \alpha(s,v)$ have a maximal element in $\mathcal{F}_0^N$ w.r.t. $\geq$? That this does not always hold for a fuzzy game is shown in Example 8.20.

Example 8.18. Let $v \in FG^{\{1,2\}}$ be given by

$$v(s_1, s_2) = \begin{cases} s_2 tg \frac{\pi s_1}{2} & \text{if } s_1 \in [0,1), \\ 0 & \text{otherwise,} \end{cases}$$

for each $s = (s_1, s_2) \in \mathcal{F}^{\{1,2\}}$. For this game $\sup_{s \in \mathcal{F}_0^{\{1,2\}}} \alpha(s, v) = \infty$.

Example 8.19. Let $v \in FG^{\{1,2,3\}}$ with

$$v(s_1, s_2, s_3) = \begin{cases} (s_1 + s_2 + s_3)^2 & \text{if } s_1, s_2, s_3 \in [0,1), \\ 0 & \text{otherwise,} \end{cases}$$

for each $s = (s_1, s_2, s_3) \in \mathcal{F}^{\{1,2,3\}}$. For this game $\sup_{s \in \mathcal{F}_0^{\{1,2,3\}}} \alpha(s, v) = 3$, and $\arg\sup_{s \in \mathcal{F}_0^{\{1,2,3\}}} \alpha(s, v) = \emptyset$.

Example 8.20. Let $v \in FG^{\{1,2\}}$ be given by

$$v(s_1, s_2) = \begin{cases} s_1 + s_2 & \text{if } s_1, s_2 \in [0,1), \\ 0 & \text{otherwise,} \end{cases}$$

for each $s = (s_1, s_2) \in \mathcal{F}^{\{1,2\}}$. For this game $\max_{s \in \mathcal{F}_0^{\{1,2\}}} \alpha(s, v) = 1$, $\arg\max_{s \in \mathcal{F}_0^{\{1,2\}}} \alpha(s, v) = [0,1) \times [0,1) \setminus \{0\}$, but this set has no maximal element w.r.t. $\geq$.

One can easily check that the games in Examples 8.18, 8.19, and 8.20 are supermodular, but not convex (CwC is not satisfied). For convex fuzzy games all three questions mentioned above are answered affirmatively in Theorem 8.23. By using this theorem, the following additional problems can also be overcome: "How to define the reduced games in the steps of the adjusted algorithm, and whether this algorithm has only a finite number of steps?"

In the proof of Lemma 8.21 we will use the notion of degree of fuzziness of a coalition (cf. page 78). Note that for $s \in \mathcal{F}_0^N$ with degree of fuzziness $\varphi(s) = 0$ we have $\alpha(s, v) \leq \max_{S \in 2^N \setminus \{\emptyset\}} \alpha(e^S, v)$, because s is equal to e^T for some $T \in 2^N \setminus \{\emptyset\}$.

Lemma 8.21. *Let $v \in CFG^N$ and $s \in \mathcal{F}_0^N$. If $\varphi(s) > 0$, then there is a $t \in \mathcal{F}_0^N$ with $\varphi(t) = \varphi(s) - 1$, $car(t) \subset car(s)$, and $\alpha(t, v) \geq \alpha(s, v)$; if $\alpha(t, v) = \alpha(s, v)$ then $t \geq s$.*

Proof. Take $s \in \mathcal{F}_0^N$ with $\varphi(s) > 0$, and $i \in N$ such that $s_i \in (0,1)$. Consider $t^0 = \left(s^{-i}, 0\right)$ and $t^1 = \left(s^{-i}, 1\right)$. Note that $\varphi\left(t^0\right) = \varphi\left(t^1\right) = \varphi(s) - 1$ and $car(t^0) \subset car(t^1) = car(s)$.

If $t^0 = e^{\emptyset}$, then $t^1 = e^i$ and then $\alpha\left(e^i, v\right) \geq \alpha\left(s_i e^i, v\right) = \alpha(s, v)$ follows from CwC. We then take $t = e^i$.

If $t^0 \neq e^{\emptyset}$ and $\alpha\left(t^0, v\right) > \alpha(s, v)$, then we take $t = t^0$.

Now, we treat the case $t^0 \neq e^{\emptyset}$ and $\alpha\left(t^0, v\right) \leq \alpha(s, v)$. From the last inequality and from the fact that $\frac{v(s)}{\lceil s \rfloor}$ is a convex combination of $\frac{v(t^0)}{\lceil t^0 \rfloor}$ and $\frac{v(s)-v(t^0)}{\lceil s-t^0 \rfloor}$, i.e.

$$\alpha(s,v) = \frac{v(s)}{\lceil s \rfloor} = \frac{\lceil t^0 \rfloor}{\lceil s \rfloor} \cdot \frac{v\left(t^0\right)}{\lceil t^0 \rfloor} + \frac{\lceil s - t^0 \rfloor}{\lceil s \rfloor} \cdot \frac{v(s) - v\left(t^0\right)}{\lceil s - t^0 \rfloor},$$

we obtain

$$\frac{v(s) - v\left(t^0\right)}{\lceil s - t^0 \rfloor} \geq \frac{v(s)}{\lceil s \rfloor} = \alpha(s,v). \tag{8.9}$$

From the fact that v satisfies IAMR (with $t^0, s, \lceil s - t^0 \rfloor, \lceil t^1 - s \rfloor$ in the roles of $s^1, s^2, \varepsilon_1, \varepsilon_2$, respectively) it follows

$$\frac{v\left(t^1\right) - v(s)}{\lceil t^1 - s \rfloor} \geq \frac{v(s) - v\left(t^0\right)}{\lceil s - t^0 \rfloor}. \tag{8.10}$$

Now, from (8.9) and (8.10) we have

$$\frac{v\left(t^1\right) - v(s)}{\lceil t^1 - s \rfloor} \geq \frac{v(s)}{\lceil s \rfloor} = \alpha(s,v). \tag{8.11}$$

Then, by applying (8.11), we obtain

$$\begin{aligned}\alpha\left(t^1, v\right) = \frac{v\left(t^1\right)}{\lceil t^1 \rfloor} &= \frac{\lceil t^1 - s \rfloor}{\lceil t^1 \rfloor} \cdot \frac{v\left(t^1\right) - v(s)}{\lceil t^1 - s \rfloor} + \frac{\lceil s \rfloor}{\lceil t^1 \rfloor} \cdot \frac{v(s)}{\lceil s \rfloor} \geq \\ &\geq \frac{\lceil t^1 - s \rfloor}{\lceil t^1 \rfloor} \cdot \frac{v(s)}{\lceil s \rfloor} + \frac{\lceil s \rfloor}{\lceil t^1 \rfloor} \cdot \frac{v(s)}{\lceil s \rfloor} = \frac{v(s)}{\lceil s \rfloor} = \alpha(s,v).\end{aligned}$$

So, we can take $t = t^1$.

From Lemma 8.21 it follows that for each $s \in \mathcal{F}_0^N$, there is a sequence $s^0, \ldots, s^k$ in $\mathcal{F}_0^N$, where $s^0 = s$ and $k = \varphi(s)$ such that $\varphi\left(s^{r+1}\right) = \varphi(s^r) - 1$, $car\left(s^{r+1}\right) \subset car\left(s^r\right)$, and $\alpha\left(s^{r+1}, v\right) \geq \alpha\left(s^r, v\right)$ for each $r \in \{0, \ldots, k-1\}$. Since $\varphi\left(s^k\right) = 0$, s^k corresponds to a crisp coalition, say T. So, we have proved

Corollary 8.22. *Let* $v \in CFG^N$. *Then for all* $s \in \mathcal{F}_0^N$ *there exists* $T \in 2^N \setminus \{\emptyset\}$ *such that* $T \subset car(s)$ *and* $\alpha\left(e^T, v\right) \geq \alpha(s, v)$.

From Corollary 8.22 it follows immediately

Theorem 8.23. *Let* $v \in CFG^N$. *Then*

(i) $\sup_{s\in\mathcal{F}_0^N} \alpha(s, v) = \max_{T\in 2^N\setminus\{\emptyset\}} \alpha\left(e^T, v\right)$;
(ii) $T^* = \max\left(\arg\max_{T\in 2^N\setminus\{\emptyset\}} \alpha\left(e^T, v\right)\right)$ *generates the largest element in* $\arg\sup_{s\in\mathcal{F}_0^N} \alpha(s, v)$, *namely* e^{T^*}.

In view of this result it is easy to adjust the Dutta-Ray algorithm to a convex fuzzy game v. In Step 1 one puts $N_1 := N$, $v_1 := v$ and considers $\arg\sup_{s\in\mathcal{F}_0^{N_1}} \alpha(s, v_1)$. According to Theorem 8.23, there is a unique maximal element in $\arg\sup_{s\in\mathcal{F}_0^{N_1}} \alpha(s, v)$, which corresponds to a crisp coalition, say S_1. Define $E_i(v) = \alpha\left(e^{S_1}, v_1\right)$ for each $i \in S_1$. If $S_1 = N$, then we stop.

In case $S_1 \neq N$, then in Step 2 one considers the convex fuzzy game v_2 with $N_2 := N_1 \setminus S_1$ and, for each $s \in [0,1]^{N\setminus S_1}$,

$$v_2(s) = v_1\left(e^{S_1} \curvearrowright s\right) - v_1\left(e^{S_1}\right),$$

where $\left(e^{S_1} \curvearrowright s\right)$ is the element in $[0,1]^N$ with

$$\left(e^{S_1} \curvearrowright s\right)_i = \begin{cases} 1 \text{ if } i \in S_1, \\ s_i \text{ if } i \in N \setminus S_1. \end{cases}$$

Once again, by using Theorem 8.23, one can take the largest element e^{S_2} in $\arg\max_{S\in 2^{N_2}\setminus\{\emptyset\}} \alpha\left(e^S, v_2\right)$ and defines $E_i(v) = \alpha\left(e^{S_2}, v_2\right)$ for all $i \in S_2$. If $S_1 \cup S_2 = N$ we stop; otherwise, we continue by considering the convex fuzzy game v_3, etc. After a finite number of steps the algorithm stops, and the obtained allocation $E(v)$ is called *the egalitarian solution of the convex fuzzy game* v.

Theorem 8.24. *Let* $v \in CFG^N$. *Then*

(i) $E(v) = E(cr(v))$;
(ii) $E(v) \in C(v)$;
(iii) $E(v)$ *Lorenz dominates every other allocation in the Aubin core* $C(v)$.

Proof. (i) This assertion follows directly from Theorem 8.23 and the adjusted Dutta-Ray algorithm given above.

(ii) Note that $E(v) = E(cr(v)) \in C(cr(v)) = C(v)$, where the first equality follows from (i), the second equality follows from Theorem 8.38(iii), and the relation $E(cr(v)) \in C(cr(v))$ is a main result in [46] for convex crisp games.

(iii) It is a fact that $E(cr(v))$ Lorenz dominates every other element of $C(cr(v))$ (cf. [46]). Since $E(v) = E(cr(v))$ and $C(cr(v)) = C(v)$, our assertion (iii) follows.

Theorem 8.24 should be interpreted as strengthening Dutta and Ray's result. One can also think that the egalitarian solution for a convex crisp game will keep the Lorenz domination property in any fuzzy extension satisfying IAMR.

The Dutta-Ray egalitarian solution for convex fuzzy games is also related to the equal division core (cf. page 88) for convex fuzzy games as Theorem 8.25 shows.

Theorem 8.25. *Let $v \in CFG^N$. Then*

(i) $E(v) \in EDC(v)$;
(ii) $EDC(v) = EDC(cr(v))$.

Proof. (i) According to Proposition 7.18(ii) and Theorem 8.24(ii), we have $E(v) \in C(v) \subseteq EDC(v)$.

(ii) The relation $EDC(v) \subset EDC(cr(v))$ follows from Proposition 7.18(i). Suppose $x \in EDC(cr(v))$. We prove that for each $s \in \mathcal{F}_0^N$ there is $i \in car(s)$ s.t. $x_i \geq \alpha(s, v)$.

Take T as in Corollary 8.22. Since $x \in EDC(cr(v))$, there is an $i \in T$ s.t. $x_i \geq \alpha(e^T, v)$. Now, from Corollary 8.22 it follows that $x_i \geq \alpha(s, v)$ for $i \in T \subset car(s)$.

The next example is meant to illustrate the various interrelations among the egalitarian solution, the core, and the equal division core for convex fuzzy games as stated in Theorems 8.24 and 8.25.

Example 8.26. Let $N = \{1, 2, 3\}$ and $T = \{1, 2\}$. Consider the unanimity fuzzy game u_{e^T} with

$$u_{e^T}(s) = \begin{cases} 1 \text{ if } s_1 = s_2 = 1, \\ 0 \text{ otherwise,} \end{cases}$$

for each $s = (s_1, s_2, s_3) \in \mathcal{F}^{\{1,2,3\}}$. According to Proposition 8.4, the game u_{e^T} is convex. Its Aubin core is given by

$$C(u_{e^T}) = co\{e^1, e^2\} = co\{(1,0,0),(0,1,0)\},$$

and the egalitarian allocation is given by

$$E(u_{e^T}) = \left(\frac{1}{2}, \frac{1}{2}, 0\right) \in C(u_{e^T}).$$

It is easy to see that $E(u_{e^T})$ Lorenz dominates every other allocation in $C(u_{e^T})$. Moreover, the equal division core $EDC(u_{e^T})$ is the set $B_1 \cup B_2$, where $B_1 = co\{e^1, \frac{1}{2}(e^1+e^2), \frac{1}{2}(e^1+e^3)\}$ and $B_2 = co\{\frac{1}{2}(e^1+e^2), e^2, \frac{1}{2}(e^2+e^3)\}$. Note that $C(u_{e^T}) \subset EDC(u_{e^T}) = EDC(cr(u_{e^T}))$.

8.3 Participation Monotonic Allocation Schemes

In this section we introduce for convex fuzzy games the notion of a *participation monotonic allocation scheme* (*pamas*). This notion is inspired by [107] where population monotonic allocation schemes (pmas) for cooperative crisp games which are necessarily totally balanced (cf. Subsections 5.1.2 and 5.2.2) are introduced. Recall that a pmas for a crisp game is a bundle of core elements, one for each subgame and the game itself, which are related via a monotonicity condition guaranteeing that each player is better off when more other players join him. In our approach (cf. [17]) the role of subgames of a crisp game will be taken over by the t-restricted games $v_t \in FG^N$ of a fuzzy game $v \in FG^N$ (cf. Definition 6.6).

Remark 8.27. Note that for each core element $x \in C(v_t)$ we have $x_i = 0$ for each $i \notin car(t)$. This follows from

$$\begin{aligned} 0 &= v(e^{\emptyset}) \\ &= v_t(e^i) \le x_i = \sum_{k\in N} x_k - \sum_{k\in N\setminus\{i\}} x_k \le v_t(e^N) - v_t(e^{N\setminus\{i\}}) = 0, \end{aligned}$$

where we use that $i \notin car(t)$ in the second and last equalities, and that $x \in C(v_t)$ in the two inequalities.

Remark 8.28. If $v \in CFG^N$, then also $v_t \in CFG^N$ for each $t \in \mathcal{F}^N$.

Definition 8.29. *Let $v \in FG^N$. A scheme $(a_{i,t})_{i\in N, t\in \mathcal{F}_0^N}$ is called a* ***participation monotonic allocation scheme*** *(pamas) if*
(i) $(a_{i,t})_{i\in N} \in C(v_t)$ for each $t \in \mathcal{F}_0^N$ (stability condition);
(ii) $t_i^{-1} a_{i,t} \geq s_i^{-1} a_{i,s}$ for each $s, t \in \mathcal{F}_0^N$ with $s \leq t$ and each $i \in car(s)$ (participation monotonicity condition).

Remark 8.30. Note that such a pamas is an $n \times \infty$-matrix, where the columns correspond to the players and the rows to the fuzzy coalitions. In each row corresponding to t there is a core element of the game v_t. The participation monotonicity condition implies that, if the scheme is used as regulator for the payoff distributions in the restricted fuzzy games, players are paid per unit of participation more in larger coalitions than in smaller coalitions.

Remark 8.31. Note that the collection of participation monotonic allocation schemes of a fuzzy game v is a convex set of $n \times \infty$-matrices.

Remark 8.32. In [120] inspired by [107], the notion of fuzzy population monotonic allocation scheme (FPMAS) is introduced. The relation between such a scheme and core elements is not studied there.

Remark 8.33. A necessary condition for the existence of a pamas for v is the existence of core elements for v_t for each $t \in \mathcal{F}_0^N$. But this is not sufficient as Example 8.34 shows. A sufficient condition is the convexity of a game as we see in Theorem 8.36.

Example 8.34. Consider the game $v \in FG^N$ with $N = \{1,2,3,4\}$ and $v(s) = \min\{s_1 + s_2, s_3 + s_4\}$ for each $s = (s_1, s_2, s_3, s_4) \in \mathcal{F}^N$. Suppose for a moment that $(a_{i,t})_{i\in N, t\in \mathcal{F}_0^N}$ is a pamas. Then for $t^1 = e^{N\backslash\{2\}}$, $t^2 = e^{N\backslash\{1\}}$, $t^3 = e^{N\backslash\{4\}}$, and $t^4 = e^{N\backslash\{3\}}$ we have $C(v_{t^k}) = \{e^k\}$, and so $\left(a_{i,t^k}\right)_{i\in N} = e^k$ for $k \in N$. But then $\sum_{k\in N} a_{k,e^N} \geq \sum_{k\in N} a_{k,t^k} = 4 > 2 = v\left(e^N\right)$, and this implies that there does not exist a pamas. Note that $C(v_t) \neq \emptyset$ holds for any $t = (t_1, t_2, t_3, t_4) \in \mathcal{F}_0^N$, because $(t_1, t_2, 0, 0) \in C(v_t)$ if $t_1 + t_2 \leq t_3 + t_4$; and $(0, 0, t_3, t_4) \in C(v_t)$ otherwise.

Definition 8.35. *Let $v \in FG^N$ and $x \in C(v)$. Then we call x* ***pamas extendable*** *if there exists a pamas $(a_{i,t})_{i\in N, t\in \mathcal{F}_0^N}$ such that $a_{i,e^N} = x_i$ for each $i \in N$.*

In the next theorem we see that convex games have a pamas. Moreover, each core element is pamas extendable.

Theorem 8.36. *Let $v \in CFG^N$ and $x \in C(v)$. Then x is pamas-extendable.*

Proof. We know from Theorem 8.38 that x is in the convex hull of the marginal vectors $m^\sigma(v)$ with $\sigma \in \pi(N)$. In view of Remark 8.31 we only need to prove that each marginal vector $m^\sigma(v)$ is pamas extendable, because then the right convex combination of these pamas extensions gives a pamas extension of x.

So, take $\sigma \in \pi(N)$ and define $(a_{i,t})_{i\in N, t\in\mathcal{F}_0^N}$ by $a_{i,t} = m_i^\sigma(v_t)$ for each $i \in N$, $t \in \mathcal{F}_0^N$. We claim that this scheme is a pamas extension of $m^\sigma(v)$.

Clearly, $a_{i,e^N} = m_i^\sigma(v)$ for each $i \in N$ since $v_{e^N} = v$. Further, by Remark 8.28, each t-restricted game v_t is a convex fuzzy game, and from Theorem 8.38 it follows that $(a_{i,t})_{i\in N} \in C(v_t)$. Hence, the scheme satisfies the stability condition.

To prove the participation monotonicity condition, take $s, t \in \mathcal{F}_0^N$ with $s \le t$ and $i \in car(s)$ and let k be the element in N such that $i = \sigma(k)$. We have to prove that $t_i^{-1}a_{i,t} \ge s_i^{-1}a_{i,s}$. Now,

$$
\begin{aligned}
t_i^{-1}a_{i,t} &= t_{\sigma(k)}^{-1} m_{\sigma(k)}^\sigma(v_t) \\
&= t_{\sigma(k)}^{-1}\left(v\left(\sum_{r=1}^{k} t_{\sigma(r)}e^{\sigma(r)}\right) - v\left(\sum_{r=1}^{k-1} t_{\sigma(r)}e^{\sigma(r)}\right)\right) \\
&\ge s_{\sigma(k)}^{-1}\left(v\left(\sum_{r=1}^{k} s_{\sigma(r)}e^{\sigma(r)}\right) - v\left(\sum_{r=1}^{k-1} s_{\sigma(r)}e^{\sigma(r)}\right)\right) \\
&= s_{\sigma(k)}^{-1} m_{\sigma(k)}^\sigma(v_s) = s_i^{-1}a_{i,s},
\end{aligned}
$$

where the inequality follows from the convexity of v (i.e. v satisfies IAMR). So $(a_{i,t})_{i\in N, t\in\mathcal{F}_0^N}$ is a pamas extension of $m^\sigma(v)$.

Further, the *total fuzzy Shapley value* of a game $v \in CFG^N$, which is the scheme $(\phi_{i,t})_{i\in N, t\in\mathcal{F}_0^N}$ with the fuzzy Shapley value of the restricted game v_t in each row corresponding to t, is a pamas. The total fuzzy Shapley value is a Shapley function (in the sense of [120]) on the class of n-person fuzzy games. For a study of a Shapley function in relation with FPMAS we refer the reader to [120].

Example 8.37. Let $v \in FG^{\{1,2\}}$ be given by $v(s_1, s_2) = 4s_1(s_1 - 2) + 10(s_2)^2$ for each $s = (s_1, s_2) \in \mathcal{F}^{\{1,2\}}$. Then v is convex and $m^{(1,2)}(v) = m^{(2,1)}(v) = \phi(v) = (-4, 10)$ because in fact v is additive: $v(s_1, s_2) = v(s_1, 0) + v(0, s_2)$. For each $t \in \mathcal{F}_0^N$ the fuzzy Shapley value

$\phi(v_t)$ equals $(4t_1(t_1-2), 10(t_2)^2)$, and the scheme $(a_{i,t})_{i\in\{1,2\},t\in\mathcal{F}_0^N}$ with $a_{1,t} = 4t_1(t_1-2)$, $a_{2,t} = 10(t_2)^2$ is a pamas extension of $\phi(v)$, with the fuzzy Shapley value of v_t in each row corresponding to t of the scheme, so $(a_{i,t})_{i\in\{1,2\},t\in\mathcal{F}_0^N}$ is the total fuzzy Shapley value of v.

8.4 Properties of Solution Concepts

This section is devoted to special properties of the solution concepts introduced so far on the class of convex fuzzy games. First, we focus on the Aubin core, the fuzzy Shapley value and the fuzzy Weber set. As we see in the following theorem the stable marginal vector property (cf. Theorem 5.10(iv)) also holds for convex fuzzy games and the fuzzy Weber set coincides with the Aubin core. Hence, the Aubin core is large; moreover it coincides with the core of the corresponding crisp game (cf. [17]).

Theorem 8.38. *Let $v \in CFG^N$. Then*

(i) $m^\sigma(v) \in C(v)$ for each $\sigma \in \pi(N)$;
(ii) $C(v) = W(v)$;
(iii) $C(v) = C(cr(v))$.

Proof. (i) For each $\sigma \in \pi(N)$ we have $\sum_{i\in N} m_i^\sigma(v) = v\left(e^N\right)$. Further, for each $\sigma \in \pi(N)$ and $s \in \mathcal{F}^N$

$$\begin{aligned}\sum_{i\in N} s_i m_i^\sigma(v) &= \sum_{k=1}^{n} s_{\sigma(k)} m_{\sigma(k)}^\sigma(v) \\ &= \sum_{k=1}^{n} s_{\sigma(k)} \left(v\left(\sum_{r=1}^{k} e^{\sigma(r)}\right) - v\left(\sum_{r=1}^{k-1} e^{\sigma(r)}\right)\right) \\ &\geq \sum_{k=1}^{n} \left(v\left(\sum_{r=1}^{k} s_{\sigma(r)} e^{\sigma(r)}\right) - v\left(\sum_{r=1}^{k-1} s_{\sigma(r)} e^{\sigma(r)}\right)\right) \\ &= v\left(\sum_{r=1}^{n} s_{\sigma(r)} e^{\sigma(r)}\right) = v(s),\end{aligned}$$

where the inequality follows by applying n times Proposition 8.8. Hence, $m^\sigma(v) \in C(v)$ for each $\sigma \in \pi(N)$.

(ii) From assertion (i) and the convexity of the core we obtain $W(v) = co\{m^\sigma(v) \mid \sigma \in \pi(N)\} \subset C(v)$. The reverse inclusion follows from Proposition 7.29.

(iii) Since $cr(v)$ is a convex crisp game by Proposition 8.5, we have $C(cr(v)) = W(cr(v))$, and $W(cr(v)) = W(v) = C(v)$ by (ii).

It follows from Theorem 8.38 that $\phi(v)$ has a central position in the Aubin core $C(v)$ if v is a convex fuzzy game. For crisp games it holds that a game v is convex if and only if $C(v) = W(v)$ (cf. Theorem 5.10(v)). For fuzzy games the implication is only in one direction. Example 8.39 presents a fuzzy game which is not convex and where the Aubin core and the fuzzy Weber set coincide.

Example 8.39. Let $v \in FG^{\{1,2\}}$ with $v(s_1, s_2) = s_1 s_2$ if $(s_1, s_2) \neq \left(\frac{1}{2}, \frac{1}{2}\right)$ and $v\left(\frac{1}{2}, \frac{1}{2}\right) = 0$. Then $v \notin CFG^{\{1,2\}}$, but $C(v) = W(v) = co\left\{(0,1), (1,0)\right\}$.

Example 8.40. Consider the public good game in Example 6.5. If the functions $g_1, \ldots, g_n$ and $-k$ are convex, then we have $v \in CFG^N$.

For fuzzy games the core is a superadditive solution, i.e.

$$C(v + w) \supset C(v) + C(w) \text{ for all } v, w \in FG^N,$$

and the fuzzy games with a non-empty Aubin core form a cone.

On the set of convex fuzzy games the Aubin core turns out to be an additive correspondence as we see in

Proposition 8.41. *The Aubin core of a convex fuzzy game and the fuzzy Shapley value are additive solutions.*

Proof. Let v, w be convex fuzzy games. Then

$$C(v + w) = C(cr(v + w)) =$$

$$C(cr(v) + cr(w)) = C(cr(v)) + C(cr(w)) = C(v) + C(w),$$

where the first equality follows from Theorem 8.38(iii) and the third equality follows from the additivity of the core for convex crisp games (cf. [25]). Further, from $\phi(v) = \phi(cr(v))$ and the additivity of the Shapley value for convex crisp games it follows that $\phi(v+w) = \phi(v)+\phi(w)$.

Now, we study properties of other cores and stable sets for convex fuzzy games (cf. [114]).

Lemma 8.42. *Let $v \in CFG^N$. Take $x, y \in I(v)$ and suppose $x \operatorname{dom}_s y$ for some $s \in \mathcal{F}_0^N$. Then $|car(s)| \geq 2$.*

Proof. Take $x, y \in I(v)$ and suppose $x \operatorname{dom}_s y$ for some $s \in \mathcal{F}_0^N$ with $car(s) = \{i\}$. Then $x_i > y_i$ and $s_i x_i \leq v(s_i e^i)$. By the convexity of v, we obtain $s_i v(e^i) \geq v(s_i e^i)$. Thus, we have $y_i < x_i \leq \frac{v(s_i e^i)}{s_i} \leq v(e^i)$ which is a contradiction with the individual rationality of y.

Theorem 8.43. *Let $v \in CFG^N$ and $w = cr(v)$. Then, for all $x, y \in I(v) = I(w)$, we have $x \operatorname{dom} y$ in v if and only if $x \operatorname{dom} y$ in w.*

Proof. First, we note that

$$I(v) = \left\{ x \in \mathbb{R}^n \mid \sum_{i \in N} x_i = v\left(e^N\right),\ x_i \geq v(e^i) \text{ for each } i \in N \right\}$$

and

$$I(w) = \left\{ x \in \mathbb{R}^n \mid \sum_{i \in N} x_i = w\left(N\right),\ x_i \geq w(\{i\}) \text{ for each } i \in N \right\}$$

coincide because $w(N) = v(e^N)$ and $w(i) = v(e^i)$ for each $i \in N$.

To prove the 'if' part, let $x, y \in I(w) = I(v)$ and $x \operatorname{dom}_S y$ for some $S \in 2^N \setminus \{\emptyset\}$. Then it implies $x \operatorname{dom}_{e^S} y$ in v.

Now, we prove the 'only if' part. Let $x, y \in I(v) = I(w)$ and $x \operatorname{dom}_s y$ for some $s \in \mathcal{F}_0^N$. Let $\varphi(s) = |\{i \in N \mid 0 < s_i < 1\}|$. By Remark 7.9, $\varphi(s) < n$. It is sufficient to prove by induction on $\varphi(s) \in \{0, \ldots, n-1\}$, that $x \operatorname{dom}_s y$ implies $x \operatorname{dom} y$ in w.

Clearly, if $\varphi(s) = 0$ then $x \operatorname{dom}_{car(s)} y$ because $\varphi(s) = 0$ implies that s is a crisp-like coalition.

Suppose now that the assertion "$x \operatorname{dom}_s y$ in v with $\varphi(s) = k$ implies $x \operatorname{dom} y$ in w" holds for each k with $0 \leq k < r < n$. Take $s \in \mathcal{F}_0^N$ with $\varphi(s) = r$, and $i \in N$ such that $0 < s_i < 1$, and take $x, y \in I(v)$ such that $x \operatorname{dom}_s y$. Then $x_i > y_i$ for each $i \in car(s)$ and $s \cdot x \leq v(s)$. Further, $|car(s)| \geq 2$ by Lemma 8.42. We note that s can be represented by a convex combination of $a = s - s_i e^i$ and $b = s + (1 - s_i)e^i$, i.e. $s = (1 - s_i)a + s_i b$. Note that $\varphi(a) = r - 1$ and $\varphi(b) = r - 1$. Further, $|car(a)| = |car(s)| - 1$, $|car(b)| = |car(s)|$.

The inequality $s \cdot x \leq v(s)$ implies $(1 - s_i)a \cdot x + s_i b \cdot x \leq v(s)$. On the other hand, the (coordinate-wise) convexity of v induces $v(s) \leq (1 - s_i)v(a) + s_i v(b)$.

Hence, $(1 - s_i)a \cdot x + s_i b \cdot x \leq v(s) \leq (1 - s_i)v(a) + s_i v(b)$ which implies $(1 - s_i)(a \cdot x - v(a)) + s_i(b \cdot x - v(b)) \leq 0$; thus $a \cdot x \leq v(a)$ or $b \cdot x \leq v(b)$. We want to show that

$$x \operatorname{dom}_a y \text{ or } x \operatorname{dom}_b y. \tag{8.12}$$

The following three cases should be considered:

(1) $b \cdot x \leq v(b)$. Then $x \operatorname{dom}_b y$, since $|car(b)| \geq 2$.
(2) $b \cdot x > v(b)$ and $|car(s)| \geq 3$. Then $a \cdot x \leq v(a)$; thus $x \operatorname{dom}_a y$, since $|car(a)| \geq 2$.
(3) $b \cdot x > v(b)$ and $|car(s)| = 2$. Then we have $a \cdot x \leq v(a)$ and $|car(b)| = 1$.

By the convexity of v and the individual rationality, we obtain $a \cdot x \geq v(a)$. In fact, let $a = s_j e^j$. Then the convexity of v induces $s_j v(e^j) \geq v(s_j e^j)$. By the individual rationality, we obtain $x_j \geq v(e^j)$. Hence, $a \cdot x = s_j x_j \geq s_j v(e^j) \geq v(s_j e^j) = v(a)$. So, $a \cdot x = v(a)$, which is contradictory to $(1 - s_i)(a \cdot x - v(a)) + s_i(b \cdot x - v(b)) \leq 0$, implying that case (3) does not occur.

Hence, (8.12) holds. Since $\varphi(a) = \varphi(b) = r - 1$, the induction hypothesis implies that $x \operatorname{dom} y$ in w.

Theorem 8.44. *Let $v \in CFG^N$. Then*
(i) $C(v) = C^P(v) = C^{cr}(v)$;
(ii) $DC(v) = DC(cr(v))$;
(iii) $C(v) = DC(v)$.

Proof. (i) For convex fuzzy games $C(v) = C(cr(v))$ (see Theorem 8.38(iii)). Now, we use Theorem 7.12(i).

(ii) From Theorem 8.43 we conclude that $DC(v) = DC(cr(v))$.

(iii) Since $v \in CFG^N$ we have $v(e^N) \geq v(s) + \sum_{i \in N \setminus car(s)} v(e^i)$ for each $s \in \mathcal{F}^N$. We obtain by Theorem 7.13 that $C^P(v) = DC(v)$. Now, we use (i).

The next theorem extends the result in [104] that each crisp convex game has a unique stable set coinciding with the dominance core.

Theorem 8.45. *Let $v \in CFG^N$. Then there is a unique stable set, namely $DC(v)$.*

Proof. By Theorem 2.12(iii), $DC(cr(v))$ is the unique stable set of $cr(v)$. In view of Theorem 8.43, the set of stable sets of v and $cr(v)$ coincide, and by Theorem 8.44(ii), we have $DC(v) = DC(cr(v))$. So, the unique stable set of v is $DC(v)$.

Note that the game v in Example 7.15 is convex, but v_1 and v_2 are not.

Remark 8.46. The relations among the Aubin core, the proper core, the crisp core, the dominance core, and the stable sets discussed above remain valid if one uses the corresponding generalized versions of these notions as presented in Section 7.3 (for details see Section 3 in [76]).

We describe now the implications of convexity on the structure of the tight catchers and compromise values introduced in Sections 7.5 and 7.6.

Let $D_k v(0)$ and $D_k v\left(e^N\right)$ for each $k \in N$ be the right and left partial derivative in the direction e^k in 0 and e^N, respectively.

Theorem 8.47. *Let $v \in CFG^N$. Then*

$$HQ(v) = \left[Dv(0), Dv\left(e^N\right)\right].$$

Proof. From the fact that v satisfies IAMR (cf. Proposition 8.8) it follows that

$$\begin{aligned} l_k\left(Q(v)\right) &= \inf\left\{\varepsilon^{-1}\left(v\left(\varepsilon e^k\right) - v(0)\right) \mid \varepsilon \in (0,1]\right\} \\ &= D_k v(0), \end{aligned}$$

and

$$\begin{aligned} u_k\left(Q(v)\right) &= \sup\left\{\varepsilon^{-1}\left(v\left(e^N\right) - v\left(e^N - \varepsilon e^k\right)\right) \mid \varepsilon \in (0,1]\right\} \\ &= D_k v\left(e^N\right). \end{aligned}$$

Theorem 8.48. *Let $v \in CFG^N$. Then $HC(v) = HW(v)$ and this hypercube is a tight catcher for $C(v) = W(v)$. Further*

$$\begin{aligned} l_k\left(C(v)\right) &= v\left(e^k\right), \\ u_k\left(C(v)\right) &= v\left(e^N\right) - v\left(e^{N\setminus\{k\}}\right) \end{aligned}$$

for each $k \in N$.

Proof. For $v \in CFG^N$ the IAMR property implies

$$\varepsilon^{-1}\left(v\left(\varepsilon e^k\right) - v(0)\right) \leq v\left(e^k\right) - v(0) \text{ for each } \varepsilon \in (0,1]$$

and

$$v\left(e^k\right) - v(0) \leq v\left(e^S + e^k\right) - v\left(e^S\right) \text{ for each } S \subset N \setminus \{k\}.$$

The first inequality corresponds to $s^1 = s^2 = 0$, $\varepsilon_1 = \varepsilon$, and $\varepsilon_2 = 1$, while the second inequality is obtained by taking $s^1 = 0$, $s^2 = e^S$, and $\varepsilon_1 = \varepsilon_2 = 1$.

So, we obtain

$$\begin{aligned} l_k\left(C(v)\right) &= \sup\left\{\varepsilon^{-1}v\left(\varepsilon e^k\right) \mid \varepsilon \in (0,1]\right\} = v\left(e^k\right) \\ &= \min\left\{v\left(e^{S\cup\{k\}}\right) - v\left(e^S\right) \mid S \subset N \setminus \{k\}\right\} \\ &= l_k\left(W(v)\right). \end{aligned}$$

Similarly, from IAMR it follows

$$\begin{aligned} u_k\left(C(v)\right) &= \inf\left\{\varepsilon^{-1}\left(v\left(e^N\right) - v\left(e^N - \varepsilon e^k\right)\right) \mid \varepsilon \in (0,1]\right\} \\ &= v\left(e^N\right) - v\left(e^{N\setminus\{k\}}\right) \\ &= \max\left\{v\left(e^{S\cup\{k\}}\right) - v\left(e^S\right) \mid S \subset N \setminus \{k\}\right\} \\ &= u_k\left(W(v)\right). \end{aligned}$$

This implies that $HC(v) = HW(v)$.

That this hypercube is a tight catcher of $C(v) = W(v)$ (see Theorem 8.38(ii)) follows from the facts that

$$\begin{aligned} l_k\left(W(v)\right) &= v\left(e^k\right) = m_k^\sigma(v), \\ u_k\left(W(v)\right) &= v\left(e^N\right) - v\left(e^{N\setminus\{k\}}\right) = m_k^\tau(v), \end{aligned}$$

where σ and τ are orderings of N with $\sigma(1) = k$ and $\tau(n) = k$, respectively.

For convex fuzzy games this theorem has consequences with respect to the coincidence of some of the compromise values introduced in Section 7.6. This can be seen in our next theorem.

Theorem 8.49. *Let $v \in CFG^N$. Then*

(i) $m_k^v\left(u\left(C(v)\right)\right) = m_k^v\left(u\left(W(v)\right)\right) = v\left(e^k\right)$ for each $k \in N$;
(ii) $val_C^\tau(v) = val_C^\sigma(v) = val_W^\tau(v) = val_W^\sigma(v)$.

Proof. (i) By Theorem 8.48, $u_k\left(C(v)\right) = u_k\left(W(v)\right) = v\left(e^N\right) - v\left(e^{N\setminus\{k\}}\right)$ for each $k \in N$. So, to prove (i), we have to show that for $k \in N$,

$$\begin{aligned} &m_k^v\left(u\left(C(v)\right)\right) \\ &= \sup\left\{s_k^{-1}\left(v(s) - \sum_{j\in N\setminus\{k\}} s_j\left(v\left(e^N\right) - v\left(e^{N\setminus\{j\}}\right)\right)\right)\right\} \\ &= v\left(e^k\right), \end{aligned}$$

where the sup is taken over $s \in \mathcal{F}^N$, $s_k > 0$.

Equivalently, we have to show that for each $s \in \mathcal{F}^N$ with $s_k > 0$

$$s_k v\left(e^k\right) \geq \sum_{j \in N \setminus \{k\}} s_j \left(v\left(e^N\right) - v\left(e^{N \setminus \{j\}}\right)\right). \tag{8.13}$$

Let σ be an ordering on N with $\sigma(1) = k$. Then

$$\begin{aligned} v(s) &= \sum_{t=1}^{n} \left(v\left(\sum_{r=1}^{t} s_{\sigma(r)} e^{\sigma(r)}\right) - v\left(\sum_{r=1}^{t-1} s_{\sigma(r)} e^{\sigma(r)}\right)\right) \\ &= v\left(s_k e^k\right) + \sum_{t=2}^{n} \left(v\left(\sum_{r=1}^{t} s_{\sigma(r)} e^{\sigma(r)}\right) - v\left(\sum_{r=1}^{t-1} s_{\sigma(r)} e^{\sigma(r)}\right)\right). \end{aligned}$$

Now, note that for each $t \in \{2, \ldots, n\}$ IAMR implies
$s_{\sigma(t)}^{-1} \left(v\left(\sum_{r=1}^{t-1} s_{\sigma(r)} e^{\sigma(r)} + s_{\sigma(t)} e^{\sigma(t)}\right) - v\left(\sum_{r=1}^{t-1} s_{\sigma(r)} e^{\sigma(r)}\right)\right)$
$\leq v\left(e^{N \setminus \{\sigma(t)\}} + 1.e^{\sigma(t)}\right) - v\left(e^{N \setminus \{\sigma(t)\}}\right)$.

So, we obtain

$$v(s) \leq s_k v\left(e^k\right) + \sum_{j \in N \setminus \{k\}} s_j \left(v\left(e^N\right) - v\left(e^{N \setminus \{j\}}\right)\right)$$

from which (8.13) follows.

(ii) Since, by (i) and Theorem 8.48, $l_k\left(C(v)\right) = m_k^v\left(u\left(C(v)\right)\right) = v\left(e^k\right)$ for each $k \in N$, it follows that $val_C^\tau(v) = val_C^\sigma(v) = val_W^\tau(v) = val_W^\sigma(v)$.

Remark 8.50. Let $v \in CFG^N$. Because $u\left(Q(v)\right) \geq u\left(C(v)\right)$, it follows from (i) in the proof of Theorem 8.49 that $m_k\left(u\left(Q(v)\right)\right) = v\left(e^k\right)$ for each $k \in N$. But in general this remainder vector is not equal to $Dv(0)$ (cf. Theorem 8.47), so in general $val_Q^\tau(v)$ and $val_Q^\sigma(v)$ do not coincide.

Example 8.51. Let $v \in CFG^{\{1,2\}}$ with $v\left(s_1, s_2\right) = s_1 \left(s_2\right)^5$ for each $s = \left(s_1, s_2\right) \in \mathcal{F}^{\{1,2\}}$. Then, by Theorems 8.38(ii) and 8.48, $C(v) = W(v) = conv\left\{m^{(1,2)}(v), m^{(2,1)}(v)\right\} = \{(0,1),(1,0)\}$ and $HC(v) = HW(v) = [(0,0),(1,1)]$. Hence, $val_C^\sigma(v) = val_W^\sigma(v) = \left(\frac{1}{2}, \frac{1}{2}\right)$. Further, $val_Q^\sigma(v) = \left(\frac{1}{6}, \frac{5}{6}\right)$ because, by Theorem 8.47, $HQ(v) = \left[Dv(e^\emptyset), Dv\left(e^{\{1,2\}}\right)\right] = [(0,0),(1,5)]$. By Theorem 8.49, $val_C^\tau(v) = val_W^\tau(v) = \left(\frac{1}{2}, \frac{1}{2}\right) = val_C^\sigma(v) = val_W^\sigma(v)$. Further, $val_Q^\tau(v)$ is the compromise between $m^v(1,5) = (0,0)$ and $(1,5)$, so in this case also $val_Q^\tau(v) = val_Q^\sigma(v) = \left(\frac{1}{6}, \frac{5}{6}\right)$.

9

Fuzzy Clan Games

In this chapter we consider fuzzy games of the form $v : [0,1]^{N_1} \times \{0,1\}^{N_2} \to \mathbb{R}$, where the players in N_1 have participation levels which may vary between 0 and 1, while the players in N_2 are crisp players in the sense that they can fully cooperate or not cooperate at all. With this kind of games we can model various economic situations where the group of agents involved is divided into two subgroups with different status: a "clan" consisting of "powerful" agents and a set of available agents willing to cooperate with the clan. This cooperation generates a positive reward only for coalitions where all clan members are present. Such situations are modeled in the classical theory of cooperative games with transferable utility by means of (total) clan games where only the full cooperation and non-cooperation at all of non-clan members with the clan are taken into account (cf. Section 5.3). Here we take over this simplifying assumption and allow non-clan members to cooperate with all clan members and some other non-clan members to a certain extent. As a result the notion of a fuzzy clan game is introduced.

9.1 The Cone of Fuzzy Clan Games

Let $N = \{1, ..., n\}$ be a finite set of players. We denote the non-empty set of clan members by C, and treat clan members as crisp players. In the following we denote the set of crisp subcoalitions of C by $\{0,1\}^C$, the set of fuzzy coalitions on $N\backslash C$ by $[0,1]^{N\backslash C}$ (equivalent to $\mathcal{F}^{N\backslash C}$), and denote $[0,1]^{N\backslash C} \times \{0,1\}^C$ by $\mathcal{F}_C^N$. For each $s \in \mathcal{F}_C^N$, $s_{N\backslash C}$ and s_C will denote its restriction to $N\backslash C$ and C, respectively. We denote the vector $(e^N)_C$ by 1_C in the following. Further we denote by $\mathcal{F}_{1_C}^N$ the set $[0,1]^{N\backslash C} \times \{1_C\}$ of fuzzy coalitions on N where all clan members have participation level 1, and where the participation level of non-clan members may vary between 0 and 1 (cf. [115]).

We define fuzzy clan games by using veto power of clan members, monotonicity, and a condition reflecting the fact that a decrease in participation level of a non-clan member in growing coalitions containing at least all clan members with full participation level results in a decrease of the average marginal return of that player (DAMR-property).

Definition 9.1. *A game $v : \mathcal{F}_C^N \to \mathbb{R}$ is a fuzzy clan game if v satisfies the following three properties:*

(i) (veto-power of clan members) $v(s) = 0$ if $s_C \neq 1_C$;
(ii) (Monotonicity) $v(s) \leq v(t)$ for all $s, t \in \mathcal{F}_C^N$ with $s \leq t$;
(iii) (DAMR-property for non-clan members) For each $i \in N \backslash C$, all $s^1, s^2 \in \mathcal{F}_{1_C}^N$ and all $\varepsilon_1, \varepsilon_2 > 0$ such that $s^1 \leq s^2$ and $0 \leq s^1 - \varepsilon_1 e^i \leq s^2 - \varepsilon_2 e^i$ we have

$$\varepsilon_1^{-1}(v(s^1) - v(s^1 - \varepsilon_1 e^i)) \geq \varepsilon_2^{-1}(v(s^2) - v(s^2 - \varepsilon_2 e^i)).$$

Property (i) expresses the fact that the full participation level of all clan members is a necessary condition for generating a positive reward for coalitions.

Fuzzy clan games for which the clan consists of a single player are called *fuzzy big boss games*, with the single clan member as the big boss.

Remark 9.2. One can see a fuzzy clan game as a special mixed action-set game, the latter being introduced in [34].

As an introduction we give two examples of interactive situations one of them leading to a fuzzy clan game, but the other one not.

Example 9.3. (A production situation with owners and gradually available workers) Let $N \backslash C = \{1, \ldots, m\}, C = \{m+1, \ldots, n\}$. Let $f : [0,1]^{N \backslash C} \to \mathbb{R}$ be a monotonic non-decreasing function with $f(0) = 0$ that satisfies the DAMR-property. Then $v : [0,1]^{N \backslash C} \times \{0,1\}^C \to \mathbb{R}$ defined by $v(s) = 0$ if $s_C \neq 1_C$ and $v(s) = f(s_1, \ldots, s_m)$ otherwise, is a fuzzy clan game with clan C. One can think of a production situation where the clan members are providers of different (complementary) essential tools needed for the production and the production function measures the gains if all clan members are cooperating with the set of workers $N \backslash C$ (cf. [89]), where each worker i can participate at level s_i which may vary from lack of participation to full participation.

Example 9.4. (A fuzzy voting situation with a fixed group with veto power) Let N and C be as in Example 9.3, and $0 < k < |N\backslash C|$. Let $v : [0,1]^{N\backslash C} \times \{0,1\}^C \to \mathbb{R}$ with

$$v(s) = \begin{cases} 1 \text{ if } s_C = 1_C \text{ and } \sum_{i=1}^m s_i \geq k, \\ 0 \text{ otherwise.} \end{cases}$$

Then v has the veto power property for members in C and the monotonicity property, but not the DAMR-property with respect to members of $N\backslash C$, hence it is not a fuzzy clan game. This game can be seen as arising from a voting situation where to pass a bill all members of C have to (fully) agree upon and the sum of the support levels $\sum_{i \in N\backslash C} s_i$ of $N \setminus C$ should exceed a fixed threshold k, where $s_i = 1$ ($s_i = 0$) correspond to full support (no support) of the bill, but also partial supports count.

In the following the set of all fuzzy clan games with a fixed nonempty set of players N and a fixed clan C is denoted by FCG_C^N. We notice that FCG_C^N is a convex cone in FG^N, that is for all $v, w \in FCG_C^N$ and $p, q \in \mathbb{R}_+$, $pv + qw \in FCG_C^N$, where $\mathbb{R}_+$ denotes the set of non-negative real numbers.

Now, we show that for each game $v \in FCG_C^N$ the corresponding crisp game $w = cr(v)$ is a total clan game if $|C| \geq 2$, and a total big boss game if $|C| = 1$.

Let $v \in FCG_C^N$. The corresponding crisp game w has the following properties which follow straightforwardly from the properties of v:

- $w(S) = 0$ if $C \not\subset S$;
- $w(S) \leq w(T)$ for all S, T with $S \subset T \subset N$;
- for all S, T with $C \subset S \subset T$ and each $i \in S \setminus C$, $w(S) - w(S\backslash\{i\}) \geq w(T) - w(T \setminus \{i\})$.

So, w is a total clan game in the terminology of [122] if $|C| \geq 2$ (cf. Subsection 5.3.2) and a total big boss game in the terminology of [27] if $|C| = 1$.

Fuzzy clan games can be seen as an extension of crisp clan games in what concerns the possibilities of cooperation available to non-clan members. Specifically, in a fuzzy clan game each non-clan member can be involved in cooperation at each extent between 0 and 1, whereas in a crisp clan game a non-clan member can only be or not a member of a (crisp) coalition containing all clan members.

In the following we consider t-restricted games corresponding to a fuzzy clan game and prove, in Proposition 9.5, that these games are also fuzzy clan games.

Let $v \in FCG_C^N$ and $t \in \mathcal{F}_{1_C}^N$. Recall that the t-restricted game v_t of v with respect to t is given by $v_t(s) = v(t * s)$ for each $s \in \mathcal{F}_C^N$.

Proposition 9.5. *Let v_t be the t-restricted game of $v \in FCG_C^N$, with $t \in \mathcal{F}_{1_C}^N$. Then $v_t \in FCG_C^N$.*

Proof. First, note that for each $s \in \mathcal{F}_C^N$ with $s_C \neq 1_C$ we have $(t*s)_C \neq 1_C$, and then the veto-power property of v implies $v_t(s) = v(t*s) = 0$. To prove the monotonicity property, let $s^1, s^2 \in \mathcal{F}_C^N$ with $s^1 \leq s^2$. Then $v_t(s^1) = v(t*s^1) \leq v(t*s^2) = v_t(s^2)$, where the inequality follows from the monotonicity of v. Now, we focus on DAMR regarding non-clan members. Let $i \in N \backslash C$, $s^1, s^2 \in \mathcal{F}_{1_C}^N$, and let $\varepsilon_1 > 0$, $\varepsilon_2 > 0$ such that $s^1 \leq s^2$ and $0 \leq s^1 - \varepsilon_1 e^i \leq s^2 - \varepsilon_2 e^i$. Then

$$\begin{aligned}\varepsilon_2^{-1}(v_t(s^2) - v_t(s^2 - \varepsilon_2 e^i)) &= \varepsilon_2^{-1}(v(t*s^2) - v(t*s^2 - t_i\varepsilon_2 e^i)) \\ &\leq \varepsilon_1^{-1}(v(t*s^1) - v(t*s^1 - t_i\varepsilon_1 e^i)) \\ &= \varepsilon_1^{-1}(v_t(s^1) - v_t(s^1 - \varepsilon_1 e^i),\end{aligned}$$

where the inequality follows from the fact that v satisfies the DAMR-property.

For each $i \in N \backslash C$, $x \in [0,1]$ and $t \in \mathcal{F}_C^N$, let $(t^{-i} \,||\, x)$ be the element in $\mathcal{F}_C^N$ such that $(t^{-i} \,||\, x)_j = t_j$ for each $j \in N \backslash \{i\}$ and $(t^{-i} \,||\, x)_i = x$. The function $v : [0,1]^{N \backslash C} \times \{0,1\}^C \to \mathbb{R}$ is called *coordinate-wise concave regarding non-clan members* if for each $i \in N \setminus C$ the function $g_{t^{-i}} : [0,1] \to \mathbb{R}$ with $g_{t^{-i}}(x) = v(t^{-i} \,||\, x)$ for each $x \in [0,1]$ is a concave function. Intuitively, this means that the function v is concave in each coordinate corresponding to (the participation level of) a non-clan member when all other coordinates are kept fixed.

The function $v : [0,1]^{N \backslash C} \times \{0,1\}^C \to \mathbb{R}$ is said to have the *submodularity property on* $[0,1]^{N \backslash C}$ if $v(s \vee t) + v(s \wedge t) \leq v(s) + v(t)$ for all $s, t \in \mathcal{F}_{1_C}^N$, where $s \vee t$ and $s \wedge t$ are those elements of $[0,1]^{N \backslash C} \times \{1_C\}$ with the i-th coordinate equal, for each $i \in N \setminus C$, to $\max\{s_i, t_i\}$ and $\min\{s_i, t_i\}$, respectively.

Remark 9.6. The DAMR-property regarding non-clan members implies two important properties of v, namely coordinate-wise concavity and submodularity. Note that the coordinate-wise concavity follows straightforwardly from the DAMR-property of v. The proof of the submodularity follows the same line as in the proof of Theorem 8.9 where it was shown that the IAMR-property implies supermodularity.

Let $\varepsilon > 0$ and $s \in \mathcal{F}_C^N$. For each $i \in N \setminus C$ we denote by $D_i v(s)$ the i-th left derivative of v in s if $s_i > 0$, and the i-th right derivative of v in s if $s_i = 0$, i.e. $D_i v(s) = \lim_{\substack{\varepsilon \to 0 \\ \varepsilon > 0}} \varepsilon^{-1}(v(s) - v(s - \varepsilon\, e^i))$, if $s_i > 0$, and $D_i v(s) = \lim_{\substack{\varepsilon \to 0 \\ \varepsilon > 0}} \varepsilon^{-1}(v(s + \varepsilon\, e^i) - v(s))$, if $s_i = 0$. It is well known that for a concave real-valued function each tangent line to the graph lies above the graph of the function. Based on this property we state

Lemma 9.7. *Let* $v \in FCG_C^N$, $t \in \mathcal{F}_{1_C}^N$, *and* $i \in N \setminus C$. *Then, for* $s_i \in [0, t_i]$, $v(t^{-i} \,\|\, t_i) - v(t^{-i} \| \, s_i) \geq (t_i - s_i) D_i v(t)$.

Proof. Applying the coordinate-wise concavity of v and the property of tangent lines to the graph of g_{-i} in $(t_i,\ g_{-i}(t_i))$ one obtains $v(t^{-i} \| \, t_i) - (t_i - s_i) D_i v(t) \geq v(t^{-i} \| \, s_i)$.

9.2 Cores and Stable Sets for Fuzzy Clan Games

We provide an explicit description of the Aubin core of a fuzzy clan game and give some insight into its geometrical shape (cf. [114] and [115]). We start with the following

Lemma 9.8. *Let* $v \in FCG_C^N$ *and* $s \in \mathcal{F}_{1_C}^N$. *Then* $v(e^N) - v(s) \geq \sum_{i \in N \setminus C} (1 - s_i) D_i v(e^N)$.

Proof. Suppose that $|N \setminus C| = m$ and denote $N \setminus C = \{1, \ldots, m\}$, $C = \{m+1, \ldots, n\}$. Let $a^0, \ldots, a^m$ and $b^1, \ldots, b^m$ be the sequences of fuzzy coalitions on N given by $a^0 = e^N$, $a^r = e^N - \sum_{k=1}^{r} (1 - s_k) e^k$, $b^r = e^N - (1 - s_r) e^r$ for each $r \in \{1, \ldots, m\}$. Note that $a^m = s \in \mathcal{F}_{1_C}^N$, and $a^{r-1} \vee b^r = e^N$, $a^{r-1} \wedge b^r = a^r$ for each $r \in \{1, \ldots, m\}$. Then

$$v(e^N) - v(s) = \sum_{r=1}^{m} \left(v\left(a^{r-1}\right) - v\left(a^r\right) \right) \geq \sum_{r=1}^{m} (v(e^N) - v(b^r)), \quad (9.1)$$

where the inequality follows from the submodularity property of v applied for each $r \in \{1, \ldots, m\}$. Now, for each $r \in \{1, \ldots, m\}$, we have by Lemma 9.7

$$D_r v(e^N) \leq (1 - s_r)^{-1} (v(e^N) - v(e^N - (1 - s_r) e^r)),$$

thus obtaining

$$v(e^N) - v(b^r) = v(e^N) - v(e^N - (1 - s_r) e^r) \geq (1 - s_r) D_r v(e^N). \quad (9.2)$$

Now, we combine (9.1) and (9.2).

Theorem 9.9. *Let* $v \in FCG_C^N$. *Then*

(i) $C(v) = \{x \in \mathbb{R}^n \mid \sum_{i=1}^n x_i = v(e^N),\ 0 \le x_i \le D_i v(e^N)$ *for each* $i \in N \setminus C,\ 0 \le x_i$ *for each* $i \in C\}$, *if* $|C| > 1$;

(ii) $C(v) = \{x \in \mathbb{R}^n \mid \sum_{i=1}^n x_i = v(e^N),\ 0 \le x_i \le D_i v(e^N)$ *for each* $i \in N \setminus \{n\},\ v(e^n) \le x_n\}$, *if* $C = \{n\}$.

Proof. We only prove (i).

(a) Let $x \in C(v)$. Then $x_i = e^i \cdot x \ge v(e^i) = 0$ for each $i \in N$ and $\sum_{i=1}^n x_i = v(e^N)$. Further, for each $i \in N \setminus C$ and each $\varepsilon \in (0, 1)$, we have

$$x_i = \varepsilon^{-1}(e^N \cdot x - (e^N - \varepsilon\, e^i) \cdot x) \le \varepsilon^{-1}(v(e^N) - v(e^N - \varepsilon\, e^i)).$$

We use now the monotonicity property and the coordinate-wise concavity property of v obtaining that $\lim_{\substack{\varepsilon \to 0 \\ \varepsilon > 0}} \varepsilon^{-1}(v(e^N) - v(e^N - \varepsilon e^i))$ exists and this limit is equal to $D_i v(e^N)$. Hence, $x_i \le D_i v(e^N)$, thus implying that $C(v)$ is a subset of the set on the right side of the equality in (i).

(b) To prove the converse inclusion, let $x \in \mathbb{R}^n$ with $\sum_{i=1}^n x_i = v(e^N)$, $0 \le x_i \le D_i v(e^N)$ for each $i \in N \setminus C$, and $0 \le x_i$ for each $i \in C$. We have to show that the inequality $s \cdot x \ge v(s)$ holds for each $s \in [0, 1]^N$. First, if $s \in [0, 1]^N$ is such that $s_C \ne 1_C$, then $v(s) = 0 \le s \cdot x$. Now, let $s \in [0, 1]^N$, with $s_C = 1_C$. Then,

$$\begin{aligned} s \cdot x = \sum_{i \in C} x_i + \sum_{i \in N \setminus C} s_i x_i &= v(e^N) - \sum_{i \in N \setminus C} (1 - s_i) x_i \\ &\ge v(e^N) - \sum_{i \in N \setminus C} (1 - s_i) D_i v(e^N). \end{aligned}$$

The inequality $s \cdot x \ge v(s)$ follows then from Lemma 9.8.

The Aubin core of a fuzzy clan game has an interesting geometric shape. It is the intersection of a simplex with "hyperbands" corresponding to the non-clan members. To be more precise, for fuzzy clan games, we have $C(v) = \Delta(v(e^N)) \cap B_1(v) \cap \cdots \cap B_m(v)$, where $\Delta(v(e^N))$ is the simplex $\{x \in \mathbb{R}_+^n \mid \sum_{i=1}^n x_i = v(e^N)\}$, and for each player $i \in \{1, \ldots, m\}$, $B_i(v) = \{x \in \mathbb{R}^n \mid 0 \le x_i \le D_i v(e^N)\}$ is the region between the two parallel hyperplanes in $\mathbb{R}^n$, $\{x \in \mathbb{R}^n \mid x_i = 0\}$ and $\{x \in \mathbb{R}^n \mid x_i = D_i v(e^N)\}$, which we call the "hyperband" corresponding to i. An interesting core element is

$$b(v) = \left(\frac{D_1 v(e^N)}{2}, \ldots, \frac{D_m v(e^N)}{2}, t, \ldots, t \right),$$

with

$$t = |C|^{-1}\left(v(e^N) - \sum_{i=1}^{m} \frac{D_i v(e^N)}{2}\right),$$

which corresponds to the point with a central location in this geometric structure. Note that $b(v)$ is in the intersection of middle-hyperplanes of all hyperbands $B_i(v)$, $i = 1, \ldots, m$, and it has the property that the coordinates corresponding to clan members are equal.

Example 9.10. For a three-person fuzzy big boss game with player 3 as the big boss and $v(e^3) = 0$ the Aubin core has the shape of a parallelogram (in the imputation set) with vertices: $(0,\ 0,\ v(e^N))$, $(D_1v(e^N),\ 0,\ v(e^N) - D_1v(e^N))$, $(0,\ D_2v(e^N),\ v(e^N) - D_2v(e^N))$, $(D_1v(e^N),\ D_2v(e^N),\ v(e^N) - D_1v(e^N) - D_2v(e^N))$. Note that

$$b(v) = \left(\frac{D_1v(e^N)}{2}, \frac{D_2v(e^N)}{2}, (e^N) - \frac{D_1v(e^N) + D_2v(e^N)}{2}\right)$$

is the middle point of this parallelogram.

For $v \in CFG^N$ we know that $C(v) = C(cr(v))$ (cf. Theorem 8.38(iii)). This is not the case in general for fuzzy clan games as the next example shows.

Example 9.11. Let $N = \{1, 2\}$, let $v : [0,1] \times \{0,1\} \to \mathbb{R}$ be given by $v(s_1, 1) = \sqrt{s_1}$, $v(s_1, 0) = 0$ for each $s_1 \in [0,1]$, and let $w = cr(v)$. Then v is a fuzzy big boss game with player 2 as the big boss, and $C(v) = \left\{(\alpha, 1-\alpha) \mid \alpha \in \left[0, \frac{1}{2}\right]\right\}$, $C(w) = \{(\alpha, 1-\alpha) \mid \alpha \in [0,1]\}$. So, $C(v) \neq C(w)$.

The next lemma plays a role in what follows.

Lemma 9.12. *Let $v \in FCG^N_C$. Let $t \in \mathcal{F}^N_{1_C}$ and v_t be the t-restricted game of v. Then, for each non-clan member $i \in car(t)$, $D_iv_t(e^N) = t_iD_iv(t)$.*

Proof. We have that

$$\begin{aligned} D_iv_t(e^N) &= \lim_{\substack{\varepsilon\to 0\\ \varepsilon>0}} \varepsilon^{-1}(v_t(e^N) - v_t(e^N - \varepsilon\, e^i)) \\ &= \lim_{\substack{\varepsilon\to 0\\ \varepsilon>0}} \varepsilon^{-1}(v(t) - v(t - \varepsilon t_i e^i) \\ &= t_iD_iv(t). \end{aligned}$$

Theorem 9.13. *Let $v \in FCG^N_C$. Then for each $t \in \mathcal{F}^N_{1_C}$ the Aubin core $C(v_t)$ of the t-restricted game v_t is described by* *(i)* $C(v_t) = \{x \in \mathbb{R}^n \mid \sum_{i \in N} x_i = v(t),\ 0 \leq x_i \leq t_i D_i v(t)$ *for each* $i \in N \setminus C,\ 0 \leq x_i$ *for each* $i \in C\}$, *if* $|C| > 1$;

(ii) $C(v_t) = \{x \in \mathbb{R}^n \mid \sum_{i \in N} x_i = v(t),\ 0 \leq x_i \leq t_i D_i v(t)$ *for each* $i \in N \setminus \{n\},\ v(t_n e^n) \leq x_n\}$, *if* $C = \{n\}$.

Proof. We only prove (i). Let $t \in \mathcal{F}^N_{1_C}$, with $|C| > 1$. Then, by the definition of the Aubin core of a fuzzy game, $C(v_t) = \{x \in \mathbb{R}^n \mid \sum_{i \in N} x_i = v_t(e^N),\ \sum_{i \in N} s_i x_i \geq v_t(s)$ for each $s \in \mathcal{F}^N_C\}$. Since $v_t(e^N) = v(t)$ and since, by Proposition 9.5, v_t is itself a fuzzy clan game, we can apply Theorem (ii)(i), thus obtaining $C(v_t) = \{x \in \mathbb{R}^n \mid \sum_{i \in N} x_i = v(t),\ 0 \leq x_i \leq D_i v_t(e^N)$ for each $i \in N \setminus C$, $0 \leq x_i$ for each $i \in C\}$. Now, we apply Lemma 9.12.

In addition to the interrelations among the different core notions and stable sets for general fuzzy games (cf. Section 7.2) the dominance core and the proper core of a fuzzy clan game coincide.

Theorem 9.14. *Let* $v \in FCG^N_C$. *Then* $DC(v) = C^P(v)$.

Proof. From the veto-power property we have that $v(e^i) = 0$ for each $i \in N$ if $|C| > 1$. Then the monotonicity of v implies $v(e^N) - \sum_{i \in N \setminus car(s)} v(e^i) - v(s) = v(e^N) - v(s) \geq 0$ for each $s \in \mathcal{F}^N$. One can easily check that in the case $|C| = 1$, $v(e^N) - \sum_{i \in N \setminus car(s)} v(e^i) - v(s) \geq 0$ for each $s \in \mathcal{F}^N$, too. The equality $DC(v) = C^P(v)$ follows then from Theorem 7.12(ii).

Now, we give two examples of fuzzy clan games v to illustrate situations in which $DC(v) \neq C(v)$ and $DC(v)$ is not a stable set, respectively.

Example 9.15. Let $N = \{1, 2\}$ and let $v : [0,1] \times \{0,1\} \to \mathbb{R}$ be given for all $s_1 \in [0,1]$ by $v(s_1, 1) = \sqrt{s_1}$ and $v(s_1, 0) = 0$. This is a big boss game with player 2 as the big boss, so $C(v) \neq \emptyset$. Moreover, as in Example 7.16, we obtain $C(v) = \left\{x \in I(v) \mid 0 \leq x_1 \leq \frac{1}{2}\right\}$, $DC(v) = \{(x_1, x_2) \in \mathbb{R}^2_+ \mid x_1 + x_2 = 1\}$, so $DC(v) \neq C(v)$. Note that $I(v) = \{(x_1, x_2) \in \mathbb{R}^2_+ \mid x_1 + x_2 = 1\}$ is the unique stable set.

The following example shows that $DC(v)$ can be a proper subset of a stable set.

Example 9.16. Let $N = \{1,2,3\}$ and v be given by $v(s_1, s_2, 0) = 0$ and $v(s_1, s_2, 1) = \min\{s_1 + s_2, 1\}$ for all $(s_1, s_2) \in [0,1]^2$. Then $DC(v) = \{(0,0,1)\}$, and no element in $I(v)$ is dominated by $(0,0,1)$. So, $DC(v)$ is not a stable set. The set $K^{a,b} = \{(\varepsilon a, \varepsilon b, 1-\varepsilon) \mid 0 \leq \varepsilon \leq 1\}$ when $a, b \in \mathbb{R}_+$ with $a + b = 1$ is a stable set of v.

9.3 Bi-Monotonic Participation Allocation Rules

We present in this section the fuzzy counterpart of a bi-monotonic allocation scheme (bi-mas) for total clan games (cf. Section 5.3.2). We call the corresponding scheme a *bi-monotonic participation allocation scheme* (bi-pamas) and study this kind of schemes with the help of a compensation-sharing rule we introduce now (cf. [115]).

Let $N \setminus C = \{1, \cdots, m\}$ and $C = \{m+1, \ldots, n\}$. We introduce for each $\alpha \in [0,1]^m$ and $\beta \in \Delta(C) = \Delta(\{m+1, \ldots, n\}) = \{z \in \mathbb{R}_+^{n-m}, \sum_{i=m+1}^{n} z_i = 1\}$ an allocation rule $\psi^{\alpha,\beta} : FCG_C^N \to \mathbb{R}^n$ whose i-th coordinate $\psi_i^{\alpha,\beta}(v)$ for each $v \in FCG_C^N$ is given by

$$\begin{cases} \alpha_i D_i v\left(e^N\right) & \text{if } i \in \{1, \ldots, m\}, \\ \beta_i \left(v\left(e^N\right) - \sum_{k=1}^{m} \alpha_k D_k v\left(e^N\right)\right) & \text{if } i \in \{m+1, \ldots, n\}. \end{cases}$$

We call this rule the *compensation-sharing rule with compensation vector α and sharing vector β*. The i-th coordinate α_i of the compensation vector α indicates that player $i \in \{1, \ldots, m\}$ obtains the part $\alpha_i D_i v(e^N)$ of his marginal contribution $D_i v(e^N)$ to e^N. Then, for each $i \in \{m+1, \ldots, n\}$, the i-th coordinate β_i of the sharing vector β determines the share

$$\beta_i \left(v\left(e^N\right) - \sum_{k=1}^{m} \alpha_k D_k v\left(e^N\right)\right)$$

for the clan member i from what is left for the group of clan members in e^N.

Theorem 9.17. *Let $v \in FCG_C^N$. Then*

(i) $\psi^{\alpha,\beta}$: $FCG_C^N \to \mathbb{R}^n$ is stable (i.e. $\psi^{\alpha,\beta}(v) \in C(v)$ for each $v \in FCG_C^N$) and additive for each $\alpha \in [0,1]^m$ and each $\beta \in \Delta(C)$;

(ii) $C(v) = \{\psi^{\alpha,\beta}(v) \mid \alpha \in [0,1]^{N \setminus C}, \beta \in \Delta(C)\}$;

(iii) the multi-function $C : FCG_C^N \to\to \mathbb{R}^n$ which assigns to each $v \in FCG_C^N$ the subset $C(v)$ of $\mathbb{R}^n$ is additive.

Proof. (i) $\psi^{\alpha,\beta}(pv + qw) = p\psi^{\alpha,\beta}(v) + q\psi^{\alpha,\beta}(w)$ for all $v, w \in FCG_C^N$ and all $p, q \in \mathbb{R}_+$, so $\psi^{\alpha,\beta}$is additive on the cone of fuzzy clan games. The stability follows from Theorem (ii).
(ii) Clearly, each $\psi^{\alpha,\beta}(v) \in C(v)$. Conversely, let $x \in C(v)$. Then, according to Theorem (ii), $x_i \in [0, D_i v(e^N)]$ for each $i \in N \setminus C$. Hence, for each $i \in \{1, \ldots, m\}$ there is $\alpha_i \in [0, 1]$ such that $x_i = \alpha_i D_i v(e^N)$. Now, we show that

$$v(e^N) - \sum_{i=1}^{m} \alpha_i D_i v(e^N) \geq 0. \tag{9.3}$$

Note that $e^C \in \mathcal{F}_{1_C}^N$ is the fuzzy coalition where each non-clan member has participation level 0 and each clan-member has participation level 1. We have

$$\begin{aligned} v(e^N) - v(e^C) &= \sum_{i=1}^{m} \left(v\left(\sum_{k=1}^{i} e^k + e^C\right) - v\left(\sum_{k=1}^{i-1} e^k + e^C\right)\right) \\ &\geq \sum_{i=1}^{m} \left(v\left(e^N\right) - v(e^N - e^i)\right) \geq \sum_{i=1}^{m} D_i v(e^N) \\ &\geq \sum_{i=1}^{m} \alpha_i D_i v(e^N), \end{aligned}$$

where the first inequality follows from the DAMR-property of v by taking $s^1 = \sum_{k=1}^{i} e^k + e^C$, $s^2 = e^N$, $\varepsilon_1 = \varepsilon_2 = 1$, the second inequality follows from Lemma 9.7 with $t = e^N$ and $s_i = 1$, and the third inequality since $D_i v(e^N) \geq 0$ in view of the monotonicity property of v. Hence, (9.3) holds.

Inequality (9.3) expresses the fact that the group of clan members is left a non-negative amount in the grand coalition.

The fact that $x_i \geq v(e^i)$ for each $i \in C$ implies that $x_i \geq 0$ for each $i \in \{m+1, \ldots, n\}$. But then there is a vector $\beta \in \Delta(C)$ such that

$$x_i = \beta_i \left(v\left(e^N\right) - \sum_{k=1}^{m} \alpha_k D_k v\left(e^N\right)\right)$$

(take $\beta \in \Delta(C)$ arbitrarily if $v(e^N) - \sum_{i=1}^{m} D_i v(e^N) = 0$, and $\beta_i = x_i \left(v\left(e^N\right) - \sum_{i=1}^{m} \alpha_i D_i v\left(e^N\right)\right)^{-1}$ for each $i \in C$, otherwise). Hence, $x = \psi^{\alpha,\beta}(v)$.

(iii) Trivially, $C(v+w) \supset C(v)+C(w)$ for all $v, w \in FCG_C^N$. Conversely, let $v, w \in FCG_C^N$. Then

$$\begin{aligned} C(v+w) &= \{\psi^{\alpha,\beta}(v+w) \mid \alpha \in [0,1]^{N\setminus C}, \beta \in \Delta(C)\} \\ &= \{\psi^{\alpha,\beta}(v) + \psi^{\alpha,\beta}(w) \mid \alpha \in [0,1]^{N\setminus C}, \beta \in \Delta(C)\} \\ &\subset \{\psi^{\alpha,\beta}(v) \mid \alpha \in [0,1]^{N\setminus C}, \beta \in \Delta(C)\} \\ &\quad + \{\psi^{\alpha,\beta}(w) \mid \alpha \in [0,1]^{N\setminus C}, \beta \in \Delta(C)\} \\ &= C(v) + C(w), \end{aligned}$$

where the equalities follow from (ii).

For fuzzy clan games the notion of bi-monotonic participation allocation scheme which we introduce now plays a similar role as pamas for convex fuzzy games (see Section 8.3).

Definition 9.18. *Let $v \in FCG_C^N$. A scheme $(b_{i,t})_{i\in N, t\in \mathcal{F}_{1_C}^N}$ is called a* ***bi-monotonic participation allocation scheme*** *(bi-pamas) for v if the following conditions hold:*
(i) (Stability) $(b_{i,t})_{i\in N} \in C(v_t)$ for each $t \in \mathcal{F}_{1_C}^N$;
(ii) (Bi-monotonicity w.r.t. participation levels) For all $s, t \in \mathcal{F}_{1_C}^N$ with $s \leq t$ we have:
(ii.1) $s_i^{-1} b_{i,s} \geq t_i^{-1} b_{i,t}$ for each $i \in (N \setminus C) \cap car(s)$;
(ii.2) $b_{i,s} \leq b_{i,t}$ for each $i \in C$.

Remark 9.19. The restriction of $(b_{i,t})_{i\in N, t\in \mathcal{F}_{1_C}^N}$ to a crisp environment (where only the crisp coalitions are considered) is a bi-monotonic allocation scheme as studied in Subsection 5.3.2.

Lemma 9.20. *Let $v \in FCG_C^N$. Let $s, t \in \mathcal{F}_{1_C}^N$ with $s \leq t$ and let $i \in car(s)$ be a non-clan member. Then $D_i v(s) \geq D_i v(t)$.*

Proof. We have that $D_i v(s) = \lim_{\substack{\varepsilon\to 0\\ \varepsilon>0}} \varepsilon^{-1}(v(s) - v(s - \varepsilon e^i)) \geq \lim_{\substack{\varepsilon\to 0\\ \varepsilon>0}} \varepsilon^{-1}(v(t) - v(t - \varepsilon\, e^i)) = D_i v(t)$, where the inequality follows from the DAMR-property of v, with $\varepsilon_1 = \varepsilon_2 = \varepsilon$.

Theorem 9.21. *Let $v \in FCG_C^N$, with $N\setminus C = \{1, \ldots, m\}$. Then for each $\alpha \in [0,1]^m$ and $\beta \in \Delta(C) = \Delta(\{m+1, \ldots, n\})$ the compensation-sharing rule $\psi^{\alpha,\beta}$ generates a bi-pamas for v, namely $\left(\psi_i^{\alpha,\beta}(v_t)\right)_{i\in N, t\in \mathcal{F}_{1_C}^N}$.*

Proof. We treat only the case $|C| > 1$. In Theorem (ii)(i) we have proved that for each $t \in \mathcal{F}_{1_C}^N$ the Aubin core $C(v_t)$ of the t-restricted game v_t is given by $C(v_t) = \{x \in \mathbb{R}^n \mid \sum_{i\in N} x_i = v(t), 0 \le x_i \le t_i D_i v(t)$ for each $i \in N \setminus C$, $0 \le x_i$ for each $i \in C\}$. Then, for each non-clan member i the α-based compensation (regardless of β) in the "grand coalition" t of the t-restricted game v_t is $\psi_i^{\alpha,\beta} = \alpha_i t_i D_i v(t)$, $i \in \{1, \ldots, m\}$. Hence, $\psi_i^{\alpha,\beta} = \beta_i \left(v\left(t\right) - \sum_{i=1}^m \alpha_i t_i D_i v\left(t\right)\right)$ for each $i \in \{m+1, \ldots, n\}$.

First, we prove that for each non-clan member i the compensation per unit of participation level is weakly decreasing when the coalition containing all clan members with full participation level and in which player i is active (i.e. $s_i > 0$) becomes larger.

Let $s, t \in \mathcal{F}_{1_C}^N$ with $s \le t$ and $i \in car(s) \cap (N \backslash C)$. We have

$$\begin{aligned}\psi_i^{\alpha,\beta}(v_s) &= \alpha_i D_i v_s(e^N) = \alpha_i s_i D_i(v_s)\\ &\ge \alpha_i s_i D_i(v_t) = \alpha_i s_i (t_i)^{-1} D_i v_t(e^N) = s_i (t_i)^{-1} \psi_i^{\alpha,\beta}(v_t),\end{aligned}$$

where the inequality follows from Lemma 9.20 and the second and third equalities by Lemma 9.12. Hence, for each $s, t \in \mathcal{F}_{1_C}^N$ with $s \le t$ and each non-clan member $i \in car(s)$

$$s_i^{-1} \psi_i^{\alpha,\beta}(v_s) \ge t_i^{-1} \psi_i^{\alpha,\beta}(v_t).$$

Now, denote by $R_\alpha(v_t)$ the α-based remainder for the clan members in the "grand coalition" t of the t-restricted game v_t. Formally,

$$R_\alpha(v_t) = v_t(e^N) - \sum_{i \in N \backslash C} \alpha_i D_i v_t(e^N) = v(t) - \sum_{i \in N \backslash C} \alpha_i D_i v(t).$$

First, we prove that for each $s, t \in \mathcal{F}_{1_C}^N$ with $s \le t$

$$R_\alpha(v_t) \ge R_\alpha(v_s). \tag{9.4}$$

Inequality (9.4) expresses the fact that the remainder for the clan members is weakly larger in larger coalitions (when non-clan members increase their participation level).

Let $s, t \in \mathcal{F}_{1_C}^N$ with $s \le t$. Then

$$\begin{aligned}v(t) - v(s) &= \sum_{k=1}^m \left(v\left(s + \sum_{i=1}^k (t_i - s_i) e^i\right) - v\left(s + \sum_{i=1}^{k-1} (t_i - s_i)\, e^i\right)\right)\\ &\ge \sum_{k=1}^m (t_k - s_k) D_k v\left(s + \sum_{i=1}^k (t_i - s_i)\, e^i\right)\end{aligned}$$

$$\geq \sum_{k=1}^{m}(t_k - s_k)D_k v(t) \geq \sum_{k=1}^{m}(t_k - s_k)\alpha_k D_k v(t),$$

where the first inequality follows from Lemma 9.7 and the second inequality from Lemma 9.20. This implies

$$\begin{aligned} v(t) - \sum_{k=1}^{m} t_k \alpha_k D_k v(t) &\geq v(s) - \sum_{k=1}^{m} s_k \alpha_k D_k v(t) \\ &\geq v(s) - \sum_{k=1}^{m} s_k \alpha_k D_k v(s), \end{aligned}$$

where the last inequality follows from Lemma 9.20. So, we proved that $R_\alpha(v_t) \geq R_\alpha(v_s)$ for all $s, t \in \mathcal{F}_{1_C}^N$ with $s \leq t$.

Now, note that inequality (9.4) implies that for each clan member the individual share (of the remainder for the whole group of clan members) in v_t, that is $\beta_i R_\alpha(v_t)$, is weakly increasing when non-clan members increase their participation level.

Definition 9.22. *Let $v \in FCG_C^N$ and $x \in C(v)$. We call x* ***bi-pamas extendable*** *if there exists a bi-pamas $(b_{i,t})_{i \in N, t \in \mathcal{F}_{1_C}^N}$ such that $b_{i,e^N} = x_i$ for each $i \in N$.*

In the next theorem we show that each core element of a fuzzy clan game is bi-pamas extendable.

Theorem 9.23. *Let $v \in FCG_C^N$ and $x \in C(v)$. Then x is bi-pamas extendable.*

Proof. Let $x \in C(v)$. Then, according to Theorem 9.17(ii), x is of the form $\psi^{\alpha,\beta}(v_{e^N})$. Take now $\left(\psi_i^{\alpha,\beta}(v_t)\right)_{i \in N, t \in \mathcal{F}_{1_C}^N}$, which is a bi-pamas by Theorem 9.21.

Part III

Multi-Choice Games

In a multi-choice game each player has a finite number of activity levels to participate with when cooperating with other players. Roughly speaking, cooperative crisp games can be seen as multi-choice games where each player has only two activity levels: full participation and no participation at all.

Multi-choice games were introduced in [60], [61] and extensively studied also in [23], [24], [28], [33], [32], [40], [53], [54], [65], [81], [82], and [88]. In this part we basically follow [23], [24], [28], and [82].

This third part of the book is organized as follows. Chapter 10 contains basic notation and notions for multi-choice games. In Chapter 11 solution concepts for multi-choice games are introduced inspired by classical solution concepts for crisp games. In Chapter 12 balanced multi-choice games, convex multi-choice games and multi-choice clan games are presented and special properties of solution concepts on these classes of games are studied.

10

Preliminaries

Let N be a non-empty finite set of players, usually of the form $\{1, \ldots, n\}$. In a multi-choice game each player $i \in N$ has a finite number of activity levels at which he or she can choose to play. In particular, any two players may have different numbers of activity levels. The reward which a group of players can obtain depends on the effort of the cooperating players. This is formalized by supposing that each player $i \in N$ has $m_i + 1$ activity levels at which he can play, where $m_i \in \mathbb{N}$. We set $M_i := \{0, \ldots, m_i\}$ as the action space of player i, where action 0 means not participating. Elements of $\mathcal{M}^N := \Pi_{i \in N} M_i$ are called *(multi-choice) coalitions*. The coalition $m = (m_1, \ldots, m_n)$ plays the role of the grand coalition. The empty coalition $(0, \ldots, 0)$ is also denoted by 0. For further use we introduce the notation $M_i^+ := M_i \setminus \{0\}$ and $\mathcal{M}_+^N := \mathcal{M}^N \setminus \{(0, \ldots, 0)\}$. For $s \in \mathcal{M}^N$ we denote by (s_{-i}, k) the participation profile where all players except player i play at levels defined by s, while player i plays at level $k \in M_i$. A particular case is $(0_{-i}, k)$, where only player i is active (at level k). For $s \in \mathcal{M}^N$ we define the carrier of s by $car(s) = \{i \in N \mid s_i > 0\}$. Let $u \in \mathcal{M}_+^N$. We denote by $\mathcal{M}_u^N$ the subset of $\mathcal{M}^N$ consisting of multi-choice coalitions $s \leq u$. For $t \in \mathcal{M}_+^N$ we also need the notation $M_i^t = \{1, ..., t_i\}$ for each $i \in N$ and $\mathcal{M}_t^N = \prod_{i \in N} M_i^t$. A characteristic function $v : \mathcal{M}^N \to \mathbb{R}$ with $v(0, \ldots, 0) = 0$ gives for each coalition $s = (s_1, \ldots, s_n) \in \mathcal{M}^N$ the worth that the players can obtain when each player i plays at level $s_i \in M_i$.

Definition 10.1. *A* ***multi-choice game*** *is a triple (N, m, v) where N is the set of players, $m \in (\mathbb{N} \cup \{0\})^N$ is the vector describing the number of activity levels for all players, and $v : \mathcal{M}^N \to \mathbb{R}$ is the characteristic function.*

If there will be no confusion, we will denote a game (N, m, v) by v. We denote the set of all multi-choice games with player set N and vector of activity levels m by $MC^{N,m}$.

Example 10.2. Consider a large building project with a deadline and a penalty for every day this deadline is exceeded. Obviously, the date of completion depends on the effort of all people involved in the project: the greater their effort the sooner the project will be completed. This situation gives rise to a multi-choice game. The worth of a coalition where each player works at a certain activity level is defined as minus the penalty that is to be paid given the completion date of the project when every player makes the corresponding effort.

Example 10.3. Suppose we are given a large company with many divisions, where the profits of the company depend on the performance of the divisions. This gives rise to a multi-choice game in which the players are the divisions and the worth of a coalition where each division functions at a certain level is the corresponding profit made by the company.

Multi-choice games can also be seen as an appropriate analytical tool for modeling cost allocation situations in which commodities are indivisible goods that are only available at a certain finite number of levels (cf. [33]).

Definition 10.4. *A game $v \in MC^{N,m}$ is called **simple** if $v(s) \in \{0,1\}$ for all $s \in \mathcal{M}^N$ and $v(m) = 1$.*

Definition 10.5. *A game $v \in MC^{N,m}$ is called **zero-normalized** if no player can gain anything by working alone, i.e. $v\left(je^i\right) = 0$ for all $i \in N$ and $j \in M_i$.*

Definition 10.6. *A game $v \in MC^{N,m}$ is called **additive** if the worth of each coalition s equals the sum of the worths of the players when they all work alone at the level they work at in s, i.e. $v(s) = \sum_{i \in N} v\left(s_i e^i\right)$ for all $s \in \mathcal{M}^N$.*

In most interesting economic applications of multi-choice games, the characteristic function v is superadditive, so that it is efficient for the players to form the grand coalition m.

Definition 10.7. *A game $v \in MC^{N,m}$ is called **superadditive** if $v(s \vee t) \geq v(s) + v(t)$ for all $s, t \in \mathcal{M}^N$ with $s \wedge t = 0$, where $(s \wedge t)_i := \min\{s_i, t_i\}$ and $(s \vee t)_i := \max\{s_i, t_i\}$ for all $i \in N$.*

Definition 10.8. *For a game $v \in MC^{N,m}$ the **zero-normalization** of v is the game v_0 that is obtained by subtracting from v the additive game a with $a\left(je^i\right) := v\left(je^i\right)$ for all $i \in N$ and $j \in M_i^+$.*

Definition 10.9. *A game $v \in MC^{N,m}$ is called **zero-monotonic** if its zero-normalization is monotonic, i.e. $v_0(s) \leq v_0(t)$ for all $s, t \in \mathcal{M}^N$ with $s \leq t$.*

Definition 10.10. *The **subgame** of $v \in MC^{N,m}$ with respect to $u \in \mathcal{M}_+^N$, v_u, is defined by $v_u(s) := v(s)$ for each $s \in \mathcal{M}_u^N$.*

Definition 10.11. *The **marginal game** of $v \in MC^{N,m}$ based on $u \in \mathcal{M}_+^N$ (or the u-marginal game of v), v^{-u}, is defined by $v^{-u}(s) := v(s+u) - v(u)$ for each $s \in \mathcal{M}_{m-u}^N$.*

The analogue of a crisp unanimity game in the multi-choice setting is the notion of a minimal effort game.

Definition 10.12. *The **minimal effort game** $u_s \in MC^{N,m}$ with $s \in \mathcal{M}_+^N$ is defined by*

$$u_s(t) := \begin{cases} 1 \text{ if } t_i \geq s_i \text{ for all } i \in N, \\ 0 \text{ otherwise}, \end{cases}$$

for all $t \in \mathcal{M}^N$.

For two sets A and B in the same vector space we set $A + B = \{x + y \mid x \in A \text{ and } y \in B\}$. By convention, the empty sum is zero.

Let $v \in MC^{N,m}$. We define $M := \{(i,j) \mid i \in N, j \in M_i\}$ and $M^+ := \left\{(i,j) \mid i \in N, j \in M_i^+\right\}$. A (level) *payoff vector* for the game v is a function $x : M \to \mathbb{R}$, where, for all $i \in N$ and $j \in M_i^+$, x_{ij} denotes the increase in payoff to player i corresponding to a change of activity level $j-1$ to j by this player, and $x_{i0} = 0$ for all $i \in N$.

Let x and y be two payoff vectors for the game v. We say that x is *weakly smaller than* y if for each $s \in \mathcal{M}^N$,

$$X(s) := \sum_{i \in N} X_{is_i} = \sum_{i \in N} \sum_{k=0}^{s_i} x_{ik} \leq \sum_{i \in N} \sum_{k=0}^{s_i} y_{ik} = \sum_{i \in N} Y_{is_i} =: Y(s).$$

Note that this does not imply that $x_{ij} \leq y_{ij}$ for all $i \in N$ and $j \in M_i$. The next example illustrates this point. To simplify the notation in the example we represent a payoff vector $x : M \to \mathbb{R}$ by a deficient matrix $[a_{ij}]$ with $i = 1, \ldots, n$ and $j = 1, \ldots, \max\{m_1, \ldots, m_n\}$. In this matrix $a_{ij} := x_{ij}$ if $i \in N$ and $j \in M_i^+$, and a_{ij} is left out $(*)$ if $i \in N$ and $j > m_i$.

Example 10.13. Let a multi-choice game be given with $N = \{1,2\}$, $m = (2,1)$ and $v((1,0)) = v((0,1)) = 1$, $v((2,0)) = 2$, $v((1,1)) = 3$ and $v((2,1)) = 5$. Consider the two payoff vectors x and y defined by

$$x = \begin{bmatrix} 1 & 2 \\ 2 & * \end{bmatrix}, \quad y = \begin{bmatrix} 2 & 1 \\ 2 & * \end{bmatrix}.$$

Then x is weakly smaller than y, since $X((1,0)) \leq Y((1,0))$, $X((1,1)) \leq Y((1,1))$ and $X(s) \leq Y(s)$ for all other s. The reason here is that player 1 gets 3 for playing at his second level according to both payoff vectors , while according to y player 1 gets 2 for playing at his first level and according to x player 1 gets only 1 at the first level.

Alternatively, we can represent a payoff vector for a game $v \in MC^{N,m}$ as a $\left(\sum_{i \in N} m_i\right)$-dimensional vector whose coordinates are numbered by corresponding elements of M^+, where the first m_1 coordinates represent payoffs for successive levels of player 1, the next m_2 coordinates are payoffs for successive levels of player 2, and so on.

Finally, we introduce different notions of marginal contributions of players to the grand coalition m in a game $v \in MC^{N,m}$.

Definition 10.14. *For each player $i \in N$, the marginal contribution of i to m in v is $w_i(m,v) = v(m) - v(m_{-i},0)$. For each $i \in N$ and $j \in \{1,\ldots,m_i - 1\}$, the marginal contribution of player i's levels which are higher than j to the grand coalition m in the game v is $w_{ij^+}(m,v) = v(m) - v(m_{-i},j)$; $w_{i0^+}(m,v) = w_i(m,v)$. For each $i \in N$ and $j \in M_i^+$, the marginal contribution of i to the coalition (m_{-i},j) is $w_{ij}(m,v) = v(m_{-i},j) - v(m_{-i},j-1)$.*

We notice that for each $i \in N$ the marginal contribution $w_i(m,v)$ of player i to m in v is equal to the sum of all marginal contributions $w_{ij}(m,v)$ of individual levels $j \in M_i^+$ of i to coalition m, that is $w_i(m,v) = \sum_{j=1}^{m_i} w_{ij}(m,v)$.

11

Solution Concepts for Multi-Choice Games

In this chapter we present extensions of solution concepts for cooperative crisp games to multi-choice games. Special attention is paid to imputations, cores and stable sets, and to solution concepts based on the marginal vectors of a multi-choice game. Shapley-like values recently introduced in the game theory literature are briefly presented.

11.1 Imputations, Cores and Stable Sets

Let $v \in MC^{N,m}$. A payoff vector $x : M \to \mathbb{R}$ is called *efficient* if $X(m) = \sum_{i\in N} \sum_{j=1}^{m_i} x_{ij} = v(m)$ and it is called *level-increase rational* if, for all $i \in N$ and $j \in M_i^+$, x_{ij} is at least the increase in worth that player i can obtain when he works alone and changes his activity from level $j-1$ to level j, i.e. $x_{ij} \geq v\left(je^i\right) - v\left((j-1)\,e^i\right)$. Notice that the level increase rationality property for (level) payoff vectors is the analogue of the individual rationality property of payoff vectors for cooperative crisp games (cf. Definition 1.27(i)).

Definition 11.1. *Let $v \in MC^{N,m}$. A payoff vector $x : M \to \mathbb{R}$ is an* ***imputation*** *of v if it is efficient and level-increase rational.*

We denote the set of imputations of a game $v \in MC^{N,m}$ by $I(v)$. It can be easily seen that

$$I(v) \neq \emptyset \Leftrightarrow \sum_{i\in N} v\left(m_i e^i\right) \leq v(m). \tag{11.1}$$

Definition 11.2. *The* ***core*** *$C(v)$ of a game $v \in MC^{N,m}$ consists of all $x \in I(v)$ that satisfy $X(s) \geq v(s)$ for all $s \in \mathcal{M}^N$.*

A notion related to the core, which is a direct generalization of the core for traditional cooperative games, was introduced in [54].

Definition 11.3. *The* ***precore*** $\mathcal{PC}(v)$ *of* $v \in MC^{N,m}$ *is defined by*

$$\mathcal{PC}(v) = \{x : M \to \mathbb{R} \mid X(m) = v(m) \text{ and } X(s) \geq v(s) \text{ for all } s \in \mathcal{M}^N\}.$$

Remark 11.4. The precore $\mathcal{PC}(v)$ of v is a convex polyhedron with infinite directions which includes the set $C(v)$. Note that precore allocations are not necessarily imputations.

Let $s \in \mathcal{M}_+^N$ and $x, y \in I(v)$. The imputation y dominates the imputation x via coalition s, denoted by $y \operatorname{dom}_s x$, if $Y(s) \leq v(s)$ and $Y_{is_i} > X_{is_i}$ for all $i \in car(s)$. We say that the imputation y dominates the imputation x if there exists $s \in \mathcal{M}_+^N$ such that $y \operatorname{dom}_s x$.

Definition 11.5. *The* ***dominance core*** $DC(v)$ *of a game* $v \in MC^{N,m}$ *consists of all* $x \in I(v)$ *for which there exists no* y *such that* y *dominates* x.

In Theorems 11.6, 11.8 and 11.9 we deal with the relations between the core and the dominance core of a multi-choice game.

Theorem 11.6. *For each game* $v \in MC^{N,m}$*, we have* $C(v) \subset DC(v)$.

Proof. Let $x \in C(v)$ and suppose $y \in I(v)$ and $s \in \mathcal{M}_+^N$, such that $y \operatorname{dom}_s x$. Then

$$v(s) \geq Y(s) = \sum_{i \in N} Y_{is_i} > \sum_{i \in N} X_{is_i} = X(s) \geq v(s),$$

which clearly gives a contradiction. Therefore, x is not dominated.

Let $v \in MC^{N,m}$ be a zero-normalized game (cf. Definition 10.8) and x a payoff vector for v. Then the condition of level increase rationality boils down to the condition $x \geq 0$. For an additive game a we have $C(a) = DC(a) = I(a) = \{x\}$, where $x : M \to \mathbb{R}$ is the payoff vector with $x_{ij} := a\left(je^i\right) - a\left((j-1)\, e^i\right)$ for all $i \in N$ and $j \in M_i^+$. Now we have the following

Proposition 11.7. *Let* $v \in MC^{N,m}$ *and* v_0 *be the zero-normalization of* v*. Let* x *be a payoff vector for* v*. Define* $y : M \to \mathbb{R}$ *by* $y_{ij} := x_{ij} - v\left(je^i\right) + v\left((j-1)\, e^i\right)$ *for all* $i \in N$ *and* $j \in M_i^+$*. Then we have*

(i) $x \in I(v) \Leftrightarrow y \in I(v_0)$,
(ii) $x \in C(v) \Leftrightarrow y \in C(v_0)$,

(iii) $x \in DC(v) \Leftrightarrow y \in DC(v_0)$.

We leave the proof of this proposition as an exercise to the reader.

Theorem 11.8. *Let $v \in MC^{N,m}$ with $DC(v) \neq \emptyset$. Then $C(v) = DC(v)$ if and only if the zero-normalization v_0 of v satisfies $v_0(s) \leq v_0(m)$ for all $s \in \mathcal{M}^N$.*

Proof. By Proposition 11.7 it suffices to prove this theorem for zero-normalized games. So, suppose v is zero-normalized. Further, suppose $C(v) = DC(v)$ and let $x \in C(v)$. Then

$$v(m) = X(m) = \sum_{i \in N} \sum_{j=1}^{s_i} x_{ij} + \sum_{i \in N} \sum_{j=s_i+1}^{m_i} x_{ij} \geq v(s)$$

for all $s \in \mathcal{M}^N$.

Now, suppose $v(s) \leq v(m)$ for all $s \in \mathcal{M}^N$. Since $C(v) \subset DC(v)$ (cf. Theorem 11.6), it suffices to prove that $x \notin DC(v)$ for all $x \in I(v) \setminus C(v)$. Let $x \in I(v) \setminus C(v)$ and $s \in \mathcal{M}_+^N$ such that $X(s) < v(s)$. Define $y : M^+ \to \mathbb{R}$ as follows

$$y_{ij} := \begin{cases} x_{ij} + \frac{v(s)-X(s)}{\sum_{k \in N} s_k} & \text{if } i \in N \text{ and } j \in \{1, \ldots, s_i\}, \\ \frac{v(m)-v(s)}{\sum_{k \in N}(m_k - s_k)} & \text{if } i \in N \text{ and } j \in \{s_i + 1, \ldots, m_i\}. \end{cases}$$

It follows readily from the definition of y that y is efficient. Since $x \geq 0$, $v(s) > X(s)$ and $v(m) \geq v(s)$, it follows that $y \geq 0$. Hence, y is also level increase rational and we conclude that $y \in I(v)$.

For $i \in N$ and $j \in \{1, \ldots, s_i\}$ we have that $y_{ij} > x_{ij}$. Hence, $Y_{is_i} > X_{is_i}$ for all $i \in N$. This and the fact that

$$Y(s) = X(s) + \sum_{i \in N} \sum_{j=1}^{s_i} \frac{v(s) - X(s)}{\sum_{k \in N} s_k} = v(s)$$

imply that $y \operatorname{dom}_s x$. Hence, $x \notin DC(v)$.

Using Theorems 11.6 and 11.8 we can easily prove Theorem 11.9. Note that this theorem also holds for cooperative crisp games, because the class of multi-choice games contains the class of cooperative crisp games.

Theorem 11.9. *Let $v \in MC^{N,m}$ with $C(v) \neq \emptyset$. Then $C(v) = DC(v)$.*

Proof. It suffices to prove the theorem for zero-normalized games (cf. Proposition 11.7). So, suppose that v is zero-normalized. From the first part of the proof of Theorem 11.8 we see that the fact that $C(v) \neq \emptyset$ implies that $v(s) \leq v(m)$ for all $s \in \mathcal{M}^N$. Because $C(v) \subset DC(v)$ (cf. Theorem 11.6), we know that $DC(v) \neq \emptyset$. Now, Theorem 11.8 immediately implies $C(v) = DC(v)$.

Considering Theorem 11.9 one might ask oneself if there actually exist games where the core is not equal to the dominance core. The answer to this question is given in Example 11.10, where we provide a multi-choice game with an empty core and a non-empty dominance core.

Example 11.10. Let a multi-choice game be given with $N = \{1,2\}$, $m = (2,1)$ and $v((1,0)) = v((0,1)) = 0$, $v((2,0)) = \frac{1}{4}$ and $v((1,1)) = v((2,1)) = 1$. An imputation x should satisfy the following (in)equalities:

$$x_{11} + x_{12} + x_{21} = 1,\ x_{11} \geq 0,\ x_{21} \geq 0,\ x_{12} \geq \frac{1}{4}.$$

Hence, we obtain

$$I(v) = co\left\{ \begin{bmatrix} 0 & \frac{1}{4} \\ \frac{3}{4} & * \end{bmatrix}, \begin{bmatrix} \frac{3}{4} & \frac{1}{4} \\ 0 & * \end{bmatrix}, \begin{bmatrix} 0 & 1 \\ 0 & * \end{bmatrix} \right\}.$$

Note that for this game an imputation can only dominate another imputation via the coalition $(1,1)$ and, since $x_{11} + x_{21} \leq \frac{3}{4}$ for all $x \in I(v)$, this gives us

$$DC(v) = co\left\{ \begin{bmatrix} 0 & \frac{1}{4} \\ \frac{3}{4} & * \end{bmatrix}, \begin{bmatrix} \frac{3}{4} & \frac{1}{4} \\ 0 & * \end{bmatrix} \right\}.$$

Finally, for none of the elements x of the dominance core $x_{11} + x_{21} \geq v((1,1))$. Since $C(v) \subset DC(v)$ one obtains $C(v) = \emptyset$. Note that for the zero-normalization v_0 of v it holds that $v_0((1,1)) = 1 > \frac{3}{4} = v_0((2,1))$.

We leave it to the reader to find an example of a cooperative crisp game for which the core is not equal to the dominance core (a game with three players will suffice).

For the game in Example 11.10 both the core and the dominance core are convex sets. This is generally true, as it is stated next.

Theorem 11.11. *Let $v \in MC^{N,m}$. Then the following two assertions hold:*

(i) $C(v)$ is convex,
(ii) $DC(v)$ is convex.

Proof. We omit the proof of part (i) because it is a simple exercise. In order to prove part (ii) it suffices to prove that $DC(v)$ is convex if v is zero-normalized. So, suppose that v is zero-normalized. Obviously, if $DC(v) = \emptyset$, then it is convex. Now, suppose $DC(v) \neq \emptyset$. We define a game $w \in MC^{N,m}$ by $w(s) := \min\{v(s), v(m)\}$ for all $s \in \mathcal{M}^N$. It can be easily seen that

$$w(m) = v(m). \tag{11.2}$$

We show that $DC(v) = DC(w) = C(w)$. Since $DC(v) \neq \emptyset$, we know that $I(v) \neq \emptyset$. Since v is zero-normalized, this implies $v(m) \geq 0$ (cf. (11.1)) and

$$w\left(je^i\right) = \min\left\{v\left(je^i\right), v(m)\right\} = 0 \tag{11.3}$$

for all $i \in N$ and $j \in M_i$.

Using (11.2) and (11.3) we see that $I(w) = I(v)$.

Now, let $s \in \mathcal{M}_+^N$ and let $x, y \in I(v) = I(w)$. Since $w(s) \leq v(s)$ we see that if $x \operatorname{dom}_s y$ in w, then $x \operatorname{dom}_s y$ in v. On the other hand, if $x \operatorname{dom}_s y$ in v, then $X(s) \leq v(s)$ and

$$X(s) = \sum_{i \in N} \sum_{j=1}^{m_i} x_{ij} - \sum_{i \in N} \sum_{j=s_i+1}^{m_i} x_{ij} \leq v(m)$$

and therefore $X(s) \leq w(s)$ and $x \operatorname{dom}_s y$ in w.

We conclude that

$$DC(w) = DC(v). \tag{11.4}$$

This implies that $DC(w) \neq \emptyset$. Since w is zero-normalized (cf. (11.3)) and

$$w(s) = \min\{v(s), v(m)\} \leq v(m) = w(m),$$

by Theorem 11.8,

$$C(w) = DC(w). \tag{11.5}$$

Now, (11.4), (11.5) and part (i) of this theorem immediately imply that $DC(v)$ is convex.

Other sets of payoff vectors for multi-choice games which are based on the notion of domination are introduced in [81] as follows.

Let $v \in MC^{N,m}$ and $2^{I(v)} := \{A \mid A \subset I(v)\}$. We introduce two maps, $D : 2^{I(v)} \to 2^{I(v)}$ and $U : 2^{I(v)} \to 2^{I(v)}$, given for all $A \subset I(v)$ by

$$\begin{aligned} D(A) &:= \{x \in I(v) \mid \text{there exists } a \in A \text{ that dominates } x\}; \\ U(A) &:= I(v) \setminus D(A). \end{aligned}$$

The set $D(A)$ consists of all imputations that are dominated by some element of A. The set $U(A)$ consists of all imputations that are undominated by elements of A. Hence, $DC(v) = U(I(v))$.

A set $A \subset I(v)$ is *internally stable* if elements of A do not dominate each other, i.e. $A \cap D(A) = \emptyset$, and it is *externally stable* if all imputations not in A are dominated by an imputation in A, i.e. $I(v) \setminus A \subset D(A)$. A set $A \subset I(v)$ is a *stable set* (cf. [78]) if it is both internally and externally stable.

It can be easily seen that a set $A \subset I(v)$ is stable if and only if A is a fixed point of U, i.e. $U(A) = A$. The following theorem is an extension towards multichoice games of Theorem 2.12.

Theorem 11.12. *Let $v \in MC^{N,m}$. Then the following two assertions hold:*

(i) Every stable set contains the dominance core as a subset;
(ii) If the dominance core is a stable set, then there are no other stable sets.

It has been shown in [68] that there exist cooperative crisp games without a stable set. Therefore, since all our definitions are consistent with the corresponding definitions for cooperative crisp games, we may conclude that multi-choice games do not always have a stable set.

In the following, we introduce the multi-choice version of the equal division core for traditional games (cf. [28]). Let $v \in MC^{N,m}$ and $s \in \mathcal{M}^N$. Let $\|s\|_1 = \sum_{i=1}^n s_i$ be the cumulate number of levels of players according to the participation profile s. Given $v \in MC^{N,m}$ and $s \in \mathcal{M}_+^N$, we denote by $\alpha(s,v)$ the (per level) average worth of s with respect to v, i.e.,

$$\alpha(s,v) := \frac{v(s)}{\|s\|_1}.$$

Note that $\alpha(s,v)$ can be interpreted as a per one-unit level increase value of coalition s.

Definition 11.13. *The* ***equal division core*** *$EDC(v)$ of $v \in MC^{N,m}$ is the set $\{x : M \to \mathbb{R} \mid X(m) = v(m);\ \nexists s \in \mathcal{M}_+^N$ s.t. $\alpha(s,v) > x_{ij}$ for all $i \in car(s),\ j \in M_i^+\}$.*

So, $x \in EDC(v)$ can be seen as a distribution of the value of the grand coalition m, where for each multi-choice coalition s, there exists a player i with a positive participation level in s and an activity level $j \in M_i^+$ such that the payoff x_{ij} is at least as good as the equal division share $\alpha(s,v)$ of $v(s)$.

The next theorem copes with the relation between the equal division core of a game $v \in MC^{N,m}$ and the precore of that game.

Theorem 11.14. *Let $v \in MC^{N,m}$. Then $\mathcal{PC}(v) \subset EDC(v)$.*

Proof. We show that $x \notin EDC(v)$ implies $x \notin \mathcal{PC}(v)$. Suppose $x \notin EDC(v)$. Then there exists $s \in \mathcal{M}_+^N$ such that $\alpha(s,v) > x_{ij}$ for all $i \in car(s)$ and $j \in \{1,...,s_i\}$. We obtain

$$X_{is_i} = \sum_{j=1}^{s_i} x_{ij} < s_i\alpha(s,v),$$

implying that

$$X(s) = \sum_{i\in N} X_{is_i} < \sum_{i\in N} s_i\alpha(s,v) = \alpha(s,v)\sum_{i\in N} s_i = v(s).$$

So, $x \notin \mathcal{PC}(v)$.

We notice that the inclusion relation in Theorem 11.14 may be strict, as in the case of cooperative crisp games.

11.2 Marginal Vectors and the Weber Set

Let $v \in MC^{N,m}$. Suppose the grand coalition $m = (m_1, \ldots, m_n)$ forms step by step, starting from the coalition $(0, \ldots, 0)$ and where in each step the level of one of the players is increased by 1. So, in particular, there are $\sum_{i\in N} m_i$ steps in this procedure. Now assign for every player to each level the marginal value that is created when the player reaches that particular level from the level directly below. This is formalized as follows.

An *admissible ordering* (for v) is a bijection $\sigma : M^+ \to \{1, \ldots, \sum_{i\in N} m_i\}$ satisfying $\sigma((i,j)) < \sigma((i,j+1))$ for all $i \in N$ and $j \in \{1, \ldots, m_i - 1\}$. The number of admissible orderings for v is $\frac{(\sum_{i\in N} m_i)!}{\Pi_{i\in N}(m_i!)}$. The set of all admissible orderings for a game v will be denoted by $\Xi(v)$.

Now let $\sigma \in \Xi(v)$ and let $k \in \{1, \ldots, \sum_{i\in N} m_i\}$. The coalition that is present after k steps according to σ, denoted by $s^{\sigma,k}$, is given by

$$s_i^{\sigma,k} := \max\left(\{j \in M_i \mid \sigma((i,j)) \leq k\} \cup \{0\}\right)$$

for all $i \in N$, and the *marginal vector* $w^\sigma : M \to \mathbb{R}$ corresponding to σ is defined by

$$w_{ij}^{\sigma} := v\left(s^{\sigma,\sigma((i,j))}\right) - v\left(s^{\sigma,\sigma((i,j))-1}\right)$$

for all $i \in N$ and $j \in M_i^+$.

In general the marginal vectors of a multi-choice game are not necessarily imputations, but for zero-monotonic games they are.

Theorem 11.15. *Let $v \in MC^{N,m}$ be zero-monotonic. Then for every $\sigma \in \Xi(v)$ the marginal vector corresponding to σ is an imputation of v.*

In the sequel we consider the convex hull of the marginal vectors of a multi-choice game, i.e., its Weber set.

Definition 11.16. *([82]) The* ***Weber set*** *$W(v)$ of a game $v \in MC^{N,m}$ is defined as*

$$W(v) := co\left\{w^{\sigma} \mid \sigma \in \Xi(v)\right\}.$$

The next theorem shows a relation between the core $C(v)$ and the Weber set $W(v)$ of a multi-choice game v.

Theorem 11.17. *Let $v \in MC^{N,m}$ and $x \in C(v)$. Then there is a $y \in W(v)$ that is weakly smaller than x.*

Proof. It will be actually proved that for each game $v \in MC^{N,m}$ and each $x \in \widetilde{C}(v)$ there is a vector $y \in W(v)$ such that y is weakly smaller than x, where $\widetilde{C}(v)$ is a core catcher of $C(v)$ $\left(C(v) \subset \widetilde{C}(v)\right)$ given by

$$\left\{x \in I(v) \mid X(s) \geq v(s) \; \forall s \in \mathcal{M}^N, x_{i0} = 0 \; \forall i \in N\right\}.$$

We will do so by induction on the number of levels involved in the game v. Two basic steps, (i) and (ii), can be distinguished.

(i) Let $v \in MC^{\{1\},m}$ with $m_1 \in \mathbb{N}$ being arbitrary. Then there is only one marginal vector y, which satisfies

$$y_{1j} = v\left(je^1\right) - v\left((j-1)\,e^1\right)$$

for all $j \in \{1, \ldots, m_1\}$. Suppose $x \in \widetilde{C}(v)$. Then

$$X\left(m_1 e^1\right) = v\left(m_1 e^1\right) = Y\left(m_1 e^1\right)$$

and

$$X\left(je^1\right) \geq v\left(je^1\right) = Y\left(je^1\right) \text{ for all } j \in \{1, \ldots, m_1\}.$$

Hence, y is weakly smaller than x.

(ii) Let $v \in MC^{\{1,2\},m}$ with $m = (1, 1)$. Then there are two marginal vectors,

$$y^1 = \begin{bmatrix} v\left(e^1\right) \\ v\left(e^1 + e^2\right) - v\left(e^1\right) \end{bmatrix} \text{ and } y^2 = \begin{bmatrix} v\left(e^1 + e^2\right) - v\left(e^2\right) \\ v\left(e^2\right) \end{bmatrix}.$$

Suppose $x \in \widetilde{C}(v)$. Then

$$x_{11} \geq v\left(e^1\right), x_{21} \geq v\left(e^2\right) \text{ and } x_{11} + x_{21} = v\left(e^1 + e^2\right).$$

Hence, x is a convex combination of y^1 and y^2. We conclude that $x \in W(v)$.

Now, we focus on the induction step. Let $v \in MC^{N,m}$ be such that $|\{i \in N \mid m_i > 0\}| \geq 2$ and $\sum_{i \in N} m_i > 2$. Suppose we already proved the statement for all games $\overline{v} \in MC^{\overline{N},\overline{m}}$ with $\sum_{i \in \overline{N}} \overline{m}_i < \sum_{i \in N} m_i$. Since, obviously, $\widetilde{C}(v)$ and $W(v)$ are both convex sets, it suffices to prove that for all extreme points x of $\widetilde{C}(v)$ we can find $y \in W(v)$ such that y is weakly smaller than x. So, let x be an extreme point of $\widetilde{C}(v)$. Then let $t \in \mathcal{M}^N$ be such that $1 \leq \sum_{i \in N} t_i \leq \sum_{i \in N} m_i - 1$ and $X(t) = v(t)$. The game v can be split up into a game u with vector of activity levels t and a game w with vector of activity levels $m - t$, defined by

$$u(s) := v(s) \text{ for all } s \in \mathcal{M}^N \text{ with } s \leq t$$

and

$$w(s) := v(s + t) - v(t) \text{ for all } s \in \mathcal{M}^N \text{ with } s \leq m - t.$$

The payoff x can be also split up into two parts,

$$x^u : \{(i,j) \mid i \in N, j \in \{0, \ldots, t_i\}\} \to \mathbb{R}$$

and

$$x^w : \{(i,j) \mid i \in N, j \in \{0, \ldots, m_i - t_i\}\} \to \mathbb{R}$$

defined by

$$x^u_{ij} := x_{ij} \text{ for all } i \in N \text{ and } j \in \{0, \ldots, t_i\}$$

and

$$x^w_{ij} := \begin{cases} x_{i,j+t_i} & \text{if } i \in N \text{ and } j \in \{1, \ldots, m_i - t_i\}, \\ 0 & \text{if } i \in N \text{ and } j = 0. \end{cases}$$

Now, $x^u \in \widetilde{C}(u)$ because $X^u(t) = X(t) = v(t) = u(t)$ and $X^u(s) = X(s) \geq v(s) = u(s)$ for all $s \in \mathcal{M}^N$ with $s \leq t$. Further, $x^w \in \widetilde{C}(w)$ because

$$\begin{aligned} X^w(m-t) &= \sum_{i\in N}\sum_{j=1}^{m_i-t_i} x_{i,j+t_i} \\ &= X(m) - X(t) = v(m) - v(t) = w(m-t) \end{aligned}$$

and

$$\begin{aligned} X^w(s) &= \sum_{i\in N}\sum_{j=1}^{s_i} x_{i,j+t_i} \\ &= X(s+t) - X(t) \geq v(s+t) - v(t) = w(s) \end{aligned}$$

for all $s \in \mathcal{M}^N$ with $s \leq m-t$.

Using the induction hypothesis, one can find $y^u \in W(u)$ such that y^u is weakly smaller than x^u, and one can find $y^w \in W(w)$ such that y^w is weakly smaller than x^w. Then $y := (y^u, y^w)$ is weakly smaller than $x := (x^u, x^w)$. Hence, the only thing to prove still is that $y \in W(v)$.

For the payoff vector

$$z^1 : \{(i,j) \mid i \in N, j \in \{0,\dots,t_i\}\} \to \mathbb{R}$$

for u and the payoff vector

$$z^2 : \{(i,j) \mid i \in N, j \in \{0,\dots,m_i - t_i\}\} \to \mathbb{R}$$

for w one defines the payoff vector $\left(z^1, z^2\right) : M \to \mathbb{R}$ for v as follows:

$$\left(z^1, z^2\right)_{ij} := \begin{cases} z^1_{ij} \text{ if } i \in N \text{ and } j \in \{0,\dots,t_i\}, \\ z^2_{ij} \text{ if } i \in N \text{ and } j \in \{t_i+1,\dots,m_i\}. \end{cases}$$

We prove that

$$(W(u), W(w)) := \left\{\left(z^1, z^2\right) \mid z^1 \in W(u), z^2 \in W(w)\right\}$$

is a subset of $W(v)$. Note that $(W(u), W(w))$ and $W(v)$ are convex sets. Hence, it suffices to prove that the extreme points of $(W(u), W(w))$ are elements of $W(v)$. Suppose $\left(z^1, z^2\right)$ is an extreme point of $(W(u), W(w))$. Then, obviously, z^1 is a marginal vector of u and z^2 is a marginal vector of w. Let $\sigma \in \Xi(u)$ and $\rho \in \Xi(w)$ be such that z^1 is the marginal vector of u corresponding to σ and z^2 is the marginal vector of w corresponding to ρ. Then $\left(z^1, z^2\right)$ is the marginal vector of v corresponding to the admissible ordering τ for v defined by $\tau((i,j)) := \sigma((i,j))$ if $i \in N$ and $j \in \{1,\dots,t_i\}$, and $\tau((i,j)) := \rho((i, j-t_i)) + \sum_{i\in N} t_i$ if $i \in N$ and $j \in \{t_i+1,\dots,m_i\}$.

Hence, $\left(z^1, z^2\right) \in W(v)$ and this completes the proof.

11.3 Shapley-like Values

It turns out that there is more than one reasonable extension of the definition of the Shapley value for cooperative crisp games to multi-choice games. First, we present the solution proposed in [82] where the average of the marginal vectors of a multi-choice game is considered to obtain an extension of the Shapley value for crisp games to multi-choice games.

Definition 11.18. *([82]) Let $v \in MC^{N,m}$. Then the* ***Shapley value*** *$\Phi(v)$ is the average of all marginal vectors of v, in formula*

$$\Phi(v) := \frac{\Pi_{i\in N}(m_i!)}{\left(\sum_{i\in N} m_i\right)!}\sum_{\sigma\in\Xi(v)} w^{\sigma}.$$

The value Φ introduced in Definition 11.18 can be characterized by additivity, the carrier property and the hierarchical strength property. The *hierarchical strength* $h_s((i,j))$ in $s \in \mathcal{M}_+^N$ of $(i,j) \in M^+$ with $j \leq s_i$ is defined by the average number of $\sigma \in \Xi(v)$ in which (i,j) is s-maximal, i.e. $h_s((i,j))$ equals

$$\frac{\Pi_{i\in N}(m_i!)}{\left(\sum_{i\in N} m_i\right)!}\left|\left\{\sigma\in\Xi(v) \mid \sigma((i,j)) = \max_{(k,l):l\leq s_k}\sigma((k,l))\right\}\right|.$$

- An allocation rule $\gamma : MC^{N,m} \to \mathbb{R}^{M^+}$ satisfies *the hierarchical strength property* if for each $v \in MC^{N,m}$ which is a multiple of a minimal effort game, say $v = \beta u_s$ with $s \in \mathcal{M}_+^N$ and $\beta \in \mathbb{R}$, we have that for all $(i_1,j_1),(i_2,j_2) \in M^+$

$$\gamma_{i_1,j_1}(v)\cdot h_s(i_2,j_2) = \gamma_{i_2,j_2}(v)\cdot h_s(i_1,j_1).$$

Next, the additivity and the carrier property of an allocation rule $\gamma : MC^{N,m} \to \mathbb{R}^{M^+}$ are introduced as follows.

- *Additivity property:* For all $v, w \in MC^{N,m}$

$$\gamma(v+w) = \gamma(v) + \gamma(w).$$

- *Carrier property:* If t is a *carrier* of $v \in MC^{N,m}$, i.e. $v(s) = v(s \wedge t)$ for all $s \in \mathcal{M}^N$, then

$$\sum_{i\in car(t)}\sum_{j=1}^{t_i}\gamma_{ij}(v) = v(m).$$

The reader is referred to [49] for the proof of the following

Theorem 11.19. *The value Φ is the unique allocation rule on $MC^{N,m}$ satisfying the additivity property, the carrier property and the hierarchical strength property.*

Now, following [81] we will focus on Shapley-like values introduced in [61]. These values were defined by using weights on the actions, thereby extending ideas of *weighted Shapley values* (cf. [63]).

When introducing the values of [61], we must restrict ourselves to multi-choice games where all players have the same number of activity levels. So, let $MC_*^{N,m}$ denote the subclass of $MC^{N,m}$ with the property that $m_i = m_j$ for all $i, j \in N$. For $v \in MC_*^{N,m}$ set $\tilde{m} := m_i$ ($i \in N$ arbitrarily) and let for each $j \in \{0, \dots, \tilde{m}\}$ a weight $w_j \in \mathbb{R}$ be associated with level j such that higher levels have larger weights, i.e. $0 = w_0 < w_1 < \dots < w_{\tilde{m}}$. The value Ψ is defined with respect to the weights w.

Definition 11.20. *([61]) For $s \in \mathcal{M}_+^N$, the value $\Psi^w(u_s)$ of the minimal effort game u_s is given by*

$$\Psi_{ij}^w(u_s) = \begin{cases} \frac{w_j}{\sum_{i \in N} w_{s_i}} & \text{if } j = s_i, \\ 0 & \text{otherwise,} \end{cases}$$

for all $i \in N$ and $j \in M_i$.

The definition of the Shapley-like value $\Psi^w(v)$ of an arbitrary game $v \in MC_*^{N,m}$ is based on the expression of v in terms of related minimal effort games (cf. Theorem 11.21) by means of the dividends of that multi-choice game. The definition of dividends for crisp games (cf. [55]) can be extended to multi-choice games as follows: for $v \in MC^{N,m}$,

$$\begin{aligned} \triangle_v(0) &:= 0 \\ \triangle_v(s) &:= v(s) - \sum_{t \le s, t \ne s} \triangle_v(t). \end{aligned} \tag{11.6}$$

Theorem 11.21. *The minimal effort games $u_s \in MC^{N,m}$, $s \in \mathcal{M}_+^N$, form a basis of the space $MC^{N,m}$. Moreover, for $v \in MC^{N,m}$ it holds that*

$$v = \sum_{s \in \mathcal{M}_+^N} \triangle_v(s) u_s.$$

Then, the value $\Psi^w(v)$ of an arbitrary game $v \in MC_*^{N,m}$ is determined by

$$\Psi^w(v) := \sum_{s \in \mathcal{M}_+^N} \Delta_v(s) \Psi^w(u_s).$$

An axiomatic characterization of this value has been provided in [61], using additivity, the carrier property, the minimal effort property and a fourth axiom that explicitly uses weights. We describe the new properties of an allocation rule $\gamma : MC^{N,m} \to \mathbb{R}^{M^+}$ below.

- *Minimal effort property:* If $v \in MC^{N,m}$ and $t \in \mathcal{M}^N$ are such that $v(s) = 0$ for all s with $s \ngeq t$, then for all $i \in N$ and $j < t_i$

$$\gamma_{ij}(v) = 0.$$

- *Weight property:* Suppose that the weights $0 = w_0 < w_1 < \ldots < w_{\tilde{m}}$ are given. If a game $v \in MC_*^{N,m}$ is a multiple of a minimal effort game, say $v = \beta u_s$, $s \in \mathcal{M}^N$, then for all $i, j \in N$

$$\gamma_{i,s_i}(v) \cdot w_{s_j} = \gamma_{j,s_j}(v) \cdot w_{s_i}.$$

The reader is referred to [61] for the proof of the following

Theorem 11.22. *Consider the class $MC_*^{N,m}$. Let weights $0 = w_0 < \ldots < w_{\tilde{m}}$ be given. Then Ψ^w is the unique allocation rule on $MC_*^{N,m}$ satisfying additivity, the carrier property, the minimal effort property and the weight property.*

Remark 11.23. The hierarchical strength property used to axiomatically characterize the value Φ introduced in Definition 11.18 incorporates the minimal effort property and the weight property, where the difference lies in the fact that the 'weights' that are used are now determined by the numbers of the activity levels of the players (cf. [49]).

An interesting question that arises now is whether the value Φ is related to the values Ψ^w. We provide an example of a multichoice game for which the value Φ is not equal to any of the values Ψ^w.

Example 11.24. Let $v \in MC^{\{1,2\},m}$ with $m = (3,3)$ and let $v = u_{(1,2)} + u_{(3,1)} + u_{(2,3)}$. There are 20 admissible orderings for this game. Some calculation shows that

$$\Phi(v) = \begin{bmatrix} \frac{4}{20} & \frac{4}{20} & \frac{19}{20} \\ \frac{1}{20} & \frac{16}{20} & \frac{16}{20} \end{bmatrix}.$$

Now, suppose we have weights $w_1 < w_2 < w_3$ associated with the activity levels. Then the corresponding value Ψ^w is

$$\Psi^w(v) = \begin{bmatrix} \frac{w_1}{w_1+w_2} & \frac{w_2}{w_2+w_3} & \frac{w_3}{w_1+w_3} \\ \frac{w_1}{w_1+w_3} & \frac{w_2}{w_1+w_2} & \frac{w_3}{w_2+w_3} \end{bmatrix}.$$

Hence, if we want to find weights w such that $\Psi^w(v) = \Phi(v)$, then these weights should satisfy the conditions $0 < w_1 < w_2 < w_3$, $w_2 = 4w_1$, $w_3 = 4w_2$ and $w_3 = 19w_1$. Clearly, it is impossible to find weights satisfying all these conditions.

The Shapley value Φ was further studied in [33] where the focus is on players' total payoffs instead of level payoff vectors. It is shown there that Φ corresponds to the discrete Aumann-Shapley method proposed in [75], and that the Aumann-Shapley value for games with a continuum of players (cf. [7]) can be obtained as the limit of multi-choice values Φ for admissible sequences of multi-choice games that converge to the continuum game.

Another extension of the Shapley value for crisp games to multi-choice games, the Shapley value Θ, was introduced in [40]. In [65] it is proved that Θ can be seen as the (level) payoff vector of average marginal contributions of the elements in $\mathcal{M}_+^N$ and it is shown via an example that in some situations Θ seems to be more appropriate than Φ. The Shapley value Θ was axiomatically characterized in [81] by extending the characterization for the Shapley value for crisp games provided in [128]. Several other characterizations of Θ were provided in [65].

A fourth Shapley-like value for multi-choice games is the egalitarian multi-choice solution ε introduced in [88]. We notice that this Shapley value makes incomplete use of information regarding the characteristic function. Specifically, the solution ε is defined using the values of those multi-choice coalitions where only one player acts at an intermediary level, while the other players are either inactive (i.e. their participation level is 0) or act at their maximum participation level. This value was axiomatically characterized in [88] by the properties of efficiency, zero contribution, additivity, anonymity, and level-symmetry.

Finally, we refer the reader to [53] for other reasonable extensions of the definition of the Shapley value for crisp games to multi-choice games.

11.4 The Equal Split-Off Set for Multi-Choice Games

In this section we focus on the multi-choice version of the equal split-off set for traditional cooperative games (cf. Section 4.2). Specifically, the equal split-off set $ESOS(v)$ of a multi-choice game $v \in MC^{N,m}$ is obtained by a procedure at each step of which one of the multi-choice coalitions with the highest (per one-unit level increase) average value is chosen and the corresponding levels divide equally the worth of that coalition. We note that such multi-choice coalitions need not be the largest coalitions with the highest average worth. Let us formally describe this algorithm which will generate multi-choice equal split-off allocations.

Step 1*:* Consider $m^1 := m$, $v_1 := v$. Select an element in $\arg\max_{s \in \mathcal{M}^N_{m^1} \setminus \{0\}} \alpha(s, v_1)$ with the maximal cumulate number of levels, say s^1. Define $e_{ij} := \alpha(s^1, v_1)$ for each $i \in car(s^1)$ and $j \in M_i^{s^1}$. If $s^1 = m$, then stop; otherwise go on.

Step k: Suppose that $s^1, s^2, ..., s^{k-1}$ have been defined recursively based on the (level) equal share principle and $s^1 + s^2 + ... + s^{k-1} \neq m$. Define a new multi-choice game with player set N and maximal participation profile $m^k := m - \sum_{i=1}^{k-1} m^i$, and $v_k(s) := v_{k-1}(s + s^{k-1}) - v_{k-1}(s^{k-1})$ for each multi-choice coalition $s \in \mathcal{M}^N_{m^k}$. Select an element s^k in $\arg\max_{s \in \mathcal{M}^N_{m^k} \setminus \{0\}} \alpha(s, v_k)$ and define $e_{ij} := \alpha(s^k, v_k)$ for all $i \in car(s^k)$ and $j \in \left\{\sum_{p=1}^{k-1} s_i^p + 1, \ldots, \sum_{p=1}^{k} s_i^p\right\}$.

In a finite number of steps, say K, where $K \leq |M^+|$, the algorithm will end, and the constructed (level) payoff vector $(e_{ij})_{(i,j) \in M^+}$ is called an *equal split-off allocation*, denoted by $e(v)$, of the multi-choice game v.

Definition 11.25. *The equal split-off set $ESOS(v)$ of $v \in MC^{N,m}$ is the set consisting of all equal split-off allocations $e(v)$.*

Remark 11.26. Note that the above described multi-choice version of the Dutta-Ray algorithm determines in K steps for each $v \in MC^{N,m}$ a sequence of (per one-unit level increase) average values $\alpha_1, \alpha_2, \ldots, \alpha_K$ with $\alpha_k := \alpha(s^k, v_k)$ for each $k \in \{1, \ldots, K\}$, and a sequence of multi-choice coalitions in M_+^N, which we denote by $t^1 := s^1$, $t^2 := s^1 + s^2$, ..., $t^k := s^1 + \ldots + s^k$, ..., $t^K := s^1 + \ldots + s^K = m$. Thus, a path $\langle t^0, t^1, \ldots, t^K \rangle$, with $t^0 = 0$ from 0 to m is obtained, to which we can associate a suitable ordered partition $T^1, T^2, \ldots, T^K$ of M, such that for all $k \in \{1, \ldots, K\}$, $T^k := \{(i,j) \mid i \in car(t^k - t^{k-1}),\ j \in$

$\{t_i^{k-1}+1,\dots,t_i^k\}$, where for each $(i,j) \in T^k$, $e_{ij} = \alpha_k$, and the coalition $t^k - t^{k-1}$ is the maximal participation profile in the "box" T^k with average worth α_k. Note that each other participation profile in T^k can be expressed as $s \wedge t^k - s \wedge t^{k-1} + t^{k-1}$, where $s \in \mathcal{M}_+^N$. Clearly, the average worth of such a participation profile is weakly smaller than α_k. Since in each step of the algorithm more than one coalition with maximal average worth might exist, each choice of such coalition will generate a suitable ordered partition. Notice that some α_k might coincide.

The next example illustrates the Dutta-Ray algorithm for multi-choice games.

Example 11.27. Consider the game $v \in MC^{\{1,2\},m}$ with $m = (2,1)$, $v(0,0) = 0$, $v(1,0) = 3$, $v(2,0) = 4$, $v(0,1) = 2$, $v(1,1) = 8$, $v(2,1) = 10$. The game is convex and we apply the above described Dutta-Ray algorithm. In Step 1, $\alpha_1 = 4$, $t^1 = s^1 = (1,1)$, and we have $e_{11} = e_{21} = 4$. In Step 2, $\alpha_2 = 2$, $t^2 = (1,1)+(1,0)$ and we have $e_{12} = 2$. We obtain $e(v) = (4,2,4)$. Note that $\alpha_1 > \alpha_2$. This is true in general, as we show in Proposition 11.28.

Proposition 11.28. *Let $v \in MC^{N,m}$ and $\alpha_k = \max_{s \in \mathcal{M}^N_{m^k} \setminus \{0\}} \frac{v_k(s)}{\|s\|_1}$ be the (level) equal share determined in Step k of the Dutta-Ray algorithm. Then $\alpha_k \geq \alpha_{k+1}$ for all $k \in \{1,\dots,K-1\}$.*

Proof. By definition of v_k and α_k, and in view of Remark 11.26, we have

$$\frac{v(t^k) - v(t^{k-1})}{\|t^k - t^{k-1}\|_1} \geq \frac{v(t^{k+1}) - v(t^{k-1})}{\|t^k - t^{k-1}\|_1 + \|t^{k+1} - t^k\|_1}.$$

By adding and subtracting $v(t^k)$ in the numerator of the right-hand term, we obtain

$$\frac{v(t^k) - v(t^{k-1})}{\|t^k - t^{k-1}\|_1} \geq \frac{v(t^{k+1}) - v(t^k) + v(t^k) - v(t^{k-1})}{\|t^k - t^{k-1}\|_1 + \|t^{k+1} - t^k\|_1}.$$

This inequality is equivalent to

$$\begin{aligned}&(v(t^k) - v(t^{k-1}))\|t^k - t^{k-1}\|_1 + (v(t^k) - v(t^{k-1}))\|t^{k+1} - t^k\|_1 \\ &\geq (v(t^{k+1}) - v(t^k))\|t^k - t^{k-1}\|_1 + (v(t^k) - v(t^{k-1}))\|t^k - t^{k-1}\|_1,\end{aligned}$$

which is, in turn, equivalent to

$$(v(t^k) - v(t^{k-1}))\|t^{k+1} - t^k\|_1 \geq (v(t^{k+1}) - v(t^k))\|t^k - t^{k-1}\|_1.$$

Remark 11.29. In a straightforward way by following the proof of Theorem 4.7 we can show that $ESOS\,(v) \subset EDC\,(v)$ for $v \in MC^{N,m}$ being superadditive.

12

Classes of Multi-Choice Games

12.1 Balanced Multi-Choice Games

12.1.1 Basic Characterizations

In [82] a notion of balancedness for multi-choice games is introduced and a theorem in the spirit of Theorem 2.4 is proved, which we present in the following.

Definition 12.1. *A game $v \in MC^{N,m}$ is called* **balanced** *if for all maps $\lambda : \mathcal{M}_+^N \to \mathbb{R}_+$ satisfying*

$$\sum_{s \in \mathcal{M}_+^N} \lambda(s) e^{car(s)} = e^N \tag{12.1}$$

it holds that $\sum_{s \in \mathcal{M}_+^N} \lambda(s) v_0(s) \leq v_0(m)$, where v_0 is the zero-normalization of v.

Note that this definition coincides with the familiar definition of balancedness for cooperative crisp games $v \in MC^{N,m}$ with $m = (1, \ldots, 1)$ (cf. Definition 1.19).

The next theorem is an extension to multi-choice games of a theorem proved in [16] and [103] which gives a necessary and sufficient condition for the nonemptiness of the core of a game.

Theorem 12.2. *Let $v \in MC^{N,m}$. Then $C(v) \neq \emptyset$ if and only if v is balanced.*

Proof. It suffices to prove the theorem for zero-normalized games.

Suppose v is zero-normalized, $C(v) \neq \emptyset$ and $x \in C(v)$. Then we define a payoff vector $y : M^+ \to \mathbb{R}$ by

$$y_{ij} := \begin{cases} 0 & \text{if } i \in N \text{ and } j \in \{2, \ldots, m_i\}, \\ \sum_{l=1}^{m_i} x_{il} & \text{if } i \in N \text{ and } j = 1. \end{cases}$$

Then, obviously, $y \in C(v)$. Further, one can identify y with the vector $(y_{11}, \ldots, y_{n1})$. This proves that $C(v) \neq \emptyset$ if and only if there exist $z_1, \ldots, z_n \in \mathbb{R}_+$ such that

$$\sum_{i \in N} z_i = v(m) \tag{12.2}$$

and

$$\sum_{i \in car(s)} z_i \geq v(s) \tag{12.3}$$

for all $s \in \mathcal{M}_0^N$.

Obviously, there exist $z_1, \ldots, z_n \in \mathbb{R}_+$ satisfying (12.2) and (12.3) if and only if for all $i \in N$ and all $s \in \mathcal{M}_+^N$ we have

$$v(m) = \min \left\{ \sum_{i \in N} z_i \mid z_i \in \mathbb{R}, \sum_{i \in car(s)} z_i \geq v(s) \right\}. \tag{12.4}$$

From the duality theorem of linear programing theory (cf. Theorem 1.33) we know that (12.4) is equivalent to

$$v(m) = \max \left\{ \sum_{s \in \mathcal{M}_+^N} \lambda(s) v(s) \mid (12.1) \text{ holds and } \lambda(s) \geq 0 \right\}. \tag{12.5}$$

It can be easily seen that (12.5) is equivalent to v being balanced.

The rest of this section deals with multi-choice flow games arising from flow situations with committee control and their relations with balanced multi-choice games. Our presentation of the results is according to [82]. Using multi-choice games to model flow situations with committee control allows one to require a coalition to make a certain effort in order to be allowed to use the corresponding arcs, for example to do a necessary amount of maintenance of the used arcs. Flow situations with committee control generate either crisp flow games when the control games on the arcs are crisp games or they generate multi-choice games when the control games on the arcs are multi-choice games. For an introduction to crisp flow games we refer the reader to [64].

Let N be a set of players and let $m \in (\mathbb{N} \cup \{0\})^N$. A flow situation consists of a directed network with two special nodes called the source and the sink. For each arc there are a capacity constraint and a constraint with respect to the allowance to use that arc. If l is an arc in the network and w is the (simple) control game for arc l, then a coalition s is allowed to use arc l if and only if $w(s) = 1$. The capacity of an arc l in the network is denoted by $c_l \in (0, \infty)$. The flow game corresponding to a flow situation assigns to a coalition s the maximal flow that coalition s can send through the network from the source to the sink.

For cooperative crisp games it was shown in [64] that a nonnegative cooperative crisp game is totally balanced if and only if it is a flow game corresponding to a flow situation in which all arcs are controlled by a single player (cf. Theorem 5.4). The corresponding definitions of a dictatorial simple game and of a totally balanced game for the multi-choice case are given below.

Definition 12.3. *A simple game $v \in MC^{N,m}$ is called* ***dictatorial*** *if there exist $i \in N$ and $j \in M_i^+$ such that $v(s) = 1$ if and only if $s_i \geq j$ for all $s \in \mathcal{M}_+^N$.*

Definition 12.4. *A game $v \in MC^{N,m}$ is called* ***totally balanced*** *if for every $s \in \mathcal{M}_+^N$ the subgame v_s is balanced, where $v_s(t) := v(t)$ for all $t \in \mathcal{M}^N$ with $t \leq s$.*

However, as exemplified in [82], one cannot generalize Theorem 5.4 to multi-choice games. In order to reach balancedness, we will restrict ourselves to zero-normalized games. Then we have the following

Theorem 12.5. *Consider a flow situation in which all control games are zero-normalized and balanced. Then the corresponding flow game $v \in MC^{N,m}$ is non-negative, zero-normalized and balanced.*

Proof. It is obvious that v is zero-normalized and non-negative. Now, in order to prove that v is balanced, let $L = \{l_1, \ldots, l_p\}$ be a set of arcs with capacities $c_1, \ldots, c_p$ and control games $w_1, \ldots, w_p$ such that every directed path from the source to the sink contains an arc in L and the capacity of L, $\sum_{r=1}^{p} c_r$, is minimal. From a theorem in [48] we find that $v(m) = \sum_{r=1}^{p} c_r$ and $v(s) \leq \sum_{r=1}^{p} c_r w_r(s)$ for all $s \in \mathcal{M}^N$.

Now, let $x^r \in C(w_r)$ for all $r \in \{1, \ldots, p\}$. Define $y := \sum_{r=1}^{p} c_r x^r$. Then

$$Y(m) = \sum_{r=1}^{p} c_r X^r(m) = \sum_{r=1}^{p} c_r w_r(m) = v(m) \tag{12.6}$$

and

$$Y(s) = \sum_{r=1}^{p} c_r X^r(s) \geq \sum_{r=1}^{p} c_r w_r(s) \geq v(m), \tag{12.7}$$

for all $s \in \mathcal{M}^N$.

Now, let $i \in N$ and $j \in M_i^+$. Since $c_r \geq 0$ and $x_{ij}^r \geq 0$ for all $r \in \{1, \ldots, p\}$ it easily follows that

$$y_{ij} = \sum_{r=1}^{p} c_r x_{ij}^r \geq 0. \tag{12.8}$$

Now (12.6), (12.7) and (12.8) imply $y \in C(v)$. Hence, v is balanced.

We can prove the converse of Theorem 12.5 using

Theorem 12.6. *Each non-negative zero-normalized balanced multi-choice game is a non-negative linear combination of zero-normalized balanced simple games.*

Proof. Let $v \in MC^{N,m}$ be non-negative, zero-normalized and balanced. We provide an algorithm to write v as a non-negative linear combination of zero-normalized balanced simple games.

Suppose $v \neq 0$ and let $x \in C(v)$. Let $k \in N$ be the smallest integer in

$$\{i \in N \mid \exists j \in N \text{ s.t. } x_{ij} > 0\}$$

and let l be the smallest integer in $\{j \in M_k^+ \mid x_{kj} > 0\}$.

Further, let

$$\beta := \min \{x_{kl}, \min \{v(s) \mid s \in \mathcal{M}_+^N, s_k \geq l, v(s) > 0\}\}$$

and let w be defined by

$$w(s) := \begin{cases} 1 \text{ if } s_k \geq l \text{ and } v(s) > 0, \\ 0 \text{ otherwise,} \end{cases}$$

for each $s \in \mathcal{M}^N$. Then w is a zero-normalized balanced simple game and $\beta > 0$.

Let $\overline{v} := v - \beta w$ and let $\overline{x} : M \to \mathbb{R}$ be defined by

$$\overline{x}_{ij} := \begin{cases} x_{kl} - \beta \text{ if } i = k \text{ and } j = l, \\ x_{ij} \qquad \text{otherwise.} \end{cases}$$

Note that $\overline{v}$ is a non-negative zero-normalized game, $v = \overline{v} + \beta w$, and $\overline{x} \in C(\overline{v})$.

Further,

$$|\{(i,j) \in M\} \mid \overline{x}_{ij} > 0| < |\{(i,j) \in M\} \mid x_{ij} > 0|$$

or

$$\left|\left\{s \in \mathcal{M}^N \mid \overline{v}(s) > 0\right\}\right| < \left|\left\{s \in \mathcal{M}^N \mid v(s) > 0\right\}\right|.$$

If $\overline{v} \neq 0$ we follow the same procedure with $\overline{v}$ in the role of v and $\overline{x}$ in the role of x. It can be easily seen that if we keep on repeating this procedure, then after finitely many steps we will obtain the zero game. Suppose this happens after q steps. Then we have found $\beta_1, \ldots, \beta_q > 0$ and zero-normalized balanced simple games $w_1, \ldots, w_q$ such that $v = \sum_{r=1}^{q} \beta_r w_r$.

Theorem 12.7. *Let $v \in MC^{N,m}$ be non-negative, zero-normalized and balanced. Then v is a flow game corresponding to a flow situation in which all control games are zero-normalized and balanced.*

Proof. According to Theorem 12.6 we can find $k \in \mathbb{N}$, $\beta_1, \ldots, \beta_k > 0$ and zero-normalized balanced games $w_1, \ldots, w_k$ such that $v = \sum_{r=1}^{k} \beta_r w_r$.

Consider now a flow situation with k arcs, where for each $r \in \{1, \ldots, k\}$ the capacity restriction of arc l_r is given by β_r and the control game of l_r is w_r. It can be easily seen that the flow game corresponding to the described flow situation is the game v.

Combining Theorems 12.5 and 12.7 we obtain

Corollary 12.8. *Let $v \in MC^{N,m}$ be non-negative and zero-normalized. Then v is balanced if and only if v is a flow game corresponding to a flow situation in which all control games are zero-normalized and balanced.*

12.1.2 Totally Balanced Games and Monotonic Allocation Schemes

The notion of level-increase monotonic allocation schemes (limas) for totally balanced multi-choice games is introduced in [28] and has been inspired by the concept of population monotonic allocation schemes for cooperative crisp games (cf. [107]).

Definition 12.9. *Let $v \in MC^{N,m}$ be totally balanced and $t \in \mathcal{M}_+^N$. We say that $a = [a_{ij}^t]_{i \in N, j \in M_i^t}^{t \in \mathcal{M}_+^N}$ is a* ***level-increase monotonic allocation scheme*** *(limas) if*

(i) $a^t \in C(v_t)$ for all $t \in \mathcal{M}_+^N$ (stability condition), and

(ii) $a_{ij}^s \leq a_{ij}^t$ *for all* $s, t \in \mathcal{M}_+^N$ *with* $s \leq t$, *for all* $i \in car(s)$ *and for all* $j \in M_i^s$ *(level-increase monotonicity condition).*

Remark 12.10. The level-increase monotonicity condition implies that, if the scheme is used as regulator for the (level) payoff distributions in the multi-choice subgames players are paid for each one-unit level increase (weakly) more in larger coalitions than in smaller coalitions.

Total balancedness of a multi-choice game is a necessary condition for the existence of a limas. A sufficient condition is the convexity of the game as we show in Section 12.2.2.

12.2 Convex Multi-Choice Games

12.2.1 Basic Characterizations

A game $v \in MC^{N,m}$ is called *convex* if

$$v(s \wedge t) + v(s \vee t) \geq v(s) + v(t) \tag{12.9}$$

for all $s, t \in \mathcal{M}^N$.

For a convex game $v \in MC^{N,m}$ it holds that

$$v(s+t) - v(s) \geq v(\overline{s}+t) - v(\overline{s}) \tag{12.10}$$

for all $s, \overline{s}, t \in \mathcal{M}^N$ satisfying $\overline{s} \leq s$, $\overline{s}_i = s_i$ for all $i \in car(t)$ and $s+t \in \mathcal{M}^N$. This can be obtained by putting s and $\overline{s}+t$ in the roles of s and t, respectively, in expression (12.9). In fact, every game satisfying expression (12.10) is convex.

Relation (12.10) can be seen as the multi-choice extension of the property of increasing marginal contributions for coalitions in traditional convex games. As a particular case, specifically when t is of the form $(0_{-i}, t_i)$ with $t_i \in M_i^+$, we obtain that the property of increasing marginal contributions for players holds true for convex multi-choice games as well, even at a more refined extent, because players can gradually increase there participation levels. For other characterizations of convex multi-choice games we refer to [28].

In the following we denote the class of convex multi-choice games with player set N and maximal participation profile m by $CMC^{N,m}$. For these games we can say more about the relation between the core and the Weber set.

Theorem 12.11. *Let* $v \in CMC^{N,m}$. *Then* $W(v) \subset C(v)$.

Proof. Note that convexity of both $C(v)$ and $W(v)$ implies that it suffices to prove that $w^\sigma \in C(v)$ for all $\sigma \in \Xi(v)$. So, let $\sigma \in \Xi(v)$. Efficiency of w^σ follows immediately from its definition. That w^σ is level increase rational follows directly when we use expression (12.10). Now, let $s \in \mathcal{M}^N$. The ordering σ induces an admissible ordering $\sigma' : \{(i,j) \mid i \in N, j \in \{1,\ldots,s_i\}\} \to \{1,\ldots,\sum_{i\in N} s_i\}$ in an obvious way. Since $s^{\sigma',\sigma'((i,j))} \leq s^{\sigma,\sigma((i,j))}$ for all $i \in N$ and $j \in \{1,\ldots,s_i\}$, the convexity of v implies $w^{\sigma'}_{ij} \leq w^{\sigma}_{ij}$ for all $i \in N$ and $j \in \{1,\ldots,s_i\}$. Hence,

$$\sum_{i\in N}\sum_{j=0}^{s_i} w^{\sigma}_{ij} \geq \sum_{i\in N}\sum_{j=0}^{s_i} w^{\sigma'}_{ij} = v(s).$$

We conclude that $w^\sigma \in C(v)$.

In contrast with convex crisp games for which $C(v) = W(v)$ holds (cf. Theorem 5.10(v)), the converse of Theorem 12.11 is not true for convex multi-choice games. We provide an example of a game $v \in CMC^{N,m}$ with $W(v) \subset C(v)$, $W(v) \neq C(v)$.

Example 12.12. Let $v \in CMC^{\{1,2\},m}$ with $m = (2,1)$ and $v((1,0)) = v((2,0)) = v((0,1)) = 0$, $v((1,1)) = 2$ and $v((2,1)) = 3$. There are three marginal vectors,

$$w_1 = \begin{bmatrix} 0 & 0 \\ 3 & * \end{bmatrix}, w_2 = \begin{bmatrix} 0 & 1 \\ 2 & * \end{bmatrix}, w_3 = \begin{bmatrix} 2 & 1 \\ 0 & * \end{bmatrix}.$$

Some calculation shows that $C(v) = co\{w_1, w_2, w_3, x\}$, where $x = \begin{bmatrix} 3 & 0 \\ 0 & * \end{bmatrix}$. We see that $x \notin co\{w_1, w_2, w_3\} = W(v)$.

The core element x in Example 12.12 seems to be too large: note that w_3 is weakly smaller than x and w_3 is still in the core $C(v)$. This inspires the following

Definition 12.13. *For a game $v \in MC^{N,m}$ the* **set** $C_{\min}(v)$ ***of minimal core elements*** *is defined as follows*

$$\{x \in C(v) \mid \nexists y \in C(v) \text{ s.t. } y \neq x \text{ and } y \text{ is weakly smaller than } x\}.$$

Now, we can formulate

Theorem 12.14. *Let $v \in CMC^{N,m}$. Then $W(v) = co(C_{\min}(v))$.*

Proof. We start by proving that all marginal vectors are minimal core elements. Let $\sigma \in \Xi(v)$. Then $w^\sigma \in C(v)$ (cf. Theorem 12.11). Suppose $y \in C(v)$ is such that $y \neq w^\sigma$ and y is weakly smaller than w^σ. Let $i \in N$ and $j \in M_i^+$ be such that $Y\left(je^i\right) < \sum_{l=1}^{j} w_{il}^\sigma$ and consider $t := s^{\sigma,\sigma((i,j))}$. Then,

$$Y(t) = \sum_{k \in N} Y\left(t_k e^k\right) < \sum_{k \in N} \sum_{l=0}^{t_k} w_{kl}^\sigma = v(t), \tag{12.11}$$

where the inequality follows from the fact that $t_i = j$ and the last equality follows from the definitions of t and w^σ. Now, (12.11) implies that $y \notin C(v)$. Hence, we see that $w^\sigma \in C_{\min}(v)$. This immediately implies that

$$W(v) \subset co\left(C_{\min}(v)\right). \tag{12.12}$$

Now, let x be a minimal core element. We prove that $x \in W(v)$. According to Theorem 11.17 we can find a payoff vector $y \in W(v)$ that is weakly smaller than x. Using (12.12) we see that $y \in co\left(C_{\min}(v)\right) \subset C(v)$. Since x is minimal we may conclude that $x = y \in W(v)$. Hence, $W(v) = co\left(C_{\min}(v)\right)$.

Note that Theorem 12.14 implies that for a convex crisp game the core coincides with the Weber set. The converse of Theorem 12.14 also holds, as shown in

Theorem 12.15. *Let $v \in MC^{N,m}$ with $W(v) = co\left(C_{\min}(v)\right)$. Then $v \in CMC^{N,m}$.*

Proof. Let $s, t \in \mathcal{M}^N$. Clearly, there is an order σ that is admissible for v and that has the property that there exist k, l with $0 \leq k \leq l \leq \sum_{i \in N} m_i$ such that $s \wedge t = s^{\sigma,k}$ and $s \vee t = s^{\sigma,l}$. Note that for the corresponding marginal vector w^σ we have that $w^\sigma \in co\left(C_{\min}(v)\right) \subset C(v)$. Using this we see

$$\begin{aligned} v(s) + v(t) &\leq \sum_{i \in N} \sum_{j=1}^{s_i} w_{ij}^\sigma + \sum_{i \in N} \sum_{j=1}^{t_i} w_{ij}^\sigma \\ &= \sum_{i \in N} \sum_{j=1}^{(s \wedge t)_i} w_{ij}^\sigma + \sum_{i \in N} \sum_{j=1}^{(s \vee t)_i} w_{ij}^\sigma \\ &= v\left(s \wedge t\right) + v\left(s \vee t\right), \end{aligned}$$

where the last equality follows from the definition of w^σ. Hence, v is convex.

From Theorems 12.14 and 12.15 we immediately obtain

Corollary 12.16. *Let $v \in MC^{N,m}$. Then $v \in CMC^{N,m}$ if and only if $W(v) = co\,(C_{\min}(v))$.*

12.2.2 Monotonic Allocation Schemes

In this section we focus on level-increase monotonic allocation schemes (limas) and show that convexity of a multi-choice game is a sufficient condition for their existence.

Definition 12.17. *Let $v \in MC^{N,m}$ and $x \in I(v)$. We say that x is limas extendable if there exists a limas $[a^t_{ij}]^{t \in \mathcal{M}^N_+}_{i \in N, j \in M^t_i}$ such that $a^m_{ij} = x_{ij}$ for each $i \in N$ and $j \in M^+_i$.*

In the next theorem we prove that each element in the convex hull of the minimal core of a convex multi-choice game is limas extendable. In the proof, restrictions of $\sigma \in \Xi(v)$ to subgames v_t of v will play a role.

Theorem 12.18. *Let $v \in CMC^{N,m}$ and $x \in co\,(C_{\min}(v))$. Then x is limas extendable.*

Proof. Let $x \in co(C_{\min}(v))$. By Theorem 12.14, x is a convex combination of marginal vectors $m^{\sigma,v}$ for $\sigma \in \Xi(v)$. Then, it suffices to prove that each marginal vector $m^{\sigma,v}$ of v is limas extendable.

Take $\sigma \in \Xi(v)$ and define $[a^t_{ij}]^{t \in \mathcal{M}^N_+}_{i \in N, j \in M^t_i}$ by $a^t_{ij} := w^{\sigma_t, v_t}_{ij}$ for each $t \in \mathcal{M}^N_+$, $i \in N$ and $j \in M^t_i$, where σ_t is the restriction of σ to t. Next, we prove that this scheme is a limas extension of $w^{\sigma,v}$.

Clearly, $a^m_{ij} = w^{\sigma,v}_{ij}$ for each $i \in N$ and $j \in M^+_i$ since $v_m = v$. Further, each multi-choice subgame v_t, $t \in \mathcal{M}^N_+$, is a convex game, and since $m^{\sigma_t, v_t} \in W(v_t) \subset C(v_t)$ (cf. Theorem 12.11), it follows that $(a^t_{ij})_{i \in N, j \in M^t_i} \in C(v_t)$. Hence, the scheme satisfies the stability condition.

To prove the participation monotonicity condition, take $s, t \in \mathcal{M}^N_+$ with $s \leq t$, $i \in car(s)$, and $j \in M^s_i \subset M^t_i$. We have to show that $a^s_{ij} \leq a^t_{ij}$. Now, $a^s_{ij} = w^{\sigma_s, v_s}_{ij} = v(u_{-i}, j) - v(u_{-i}, j-1)$, where (u_{-i}, j) is the intermediary multi-choice coalition in the maximal chain generated by the restriction of σ to s, when player i increased his participation level from $j-1$ to j. Similarly, $a^t_{ij} = w^{\sigma_t, v_t}_{ij} = v(\bar{u}_{-i}, j) - v(\bar{u}_{-i}, j-1)$.

Note that, since $s \leq t$, in the maximal chain generated by σ_s the turn of i to increase his participation level from $j-1$ to j will come not later than the same turn in the maximal chain generated by σ_t, implying that $(u_{-i}, j) \leq (\bar{u}_{-i}, j)$. Furthermore, $(\bar{u}_{-i}, j) \leq m$. Then,

$$a_{ij}^s = v(u_{-i}, j) - v(u_{-i}, j-1) \leq v(\bar{u}_{-i}, j) - v(\bar{u}_{-i}, j-1) = a_{ij}^t,$$

where the inequality follows from the convexity of v.

Specifically, we used relation (12.10) with $(u_{-i}, j-1)$ in the role of $\bar{s}$, $(\overline{u}_{-i}, j-1)$ in the role of s, and $(0_{-i}, 1)$ in the role of t. Hence, $[a_{ij}^t]_{i \in N, j \in M_i^t}^{t \in \mathcal{M}_+^N}$ is a limas extension of $w^{\sigma,v}$.

Since the Shapley value (cf. Definition 11.18) of a convex multi-choice game is the average of the marginal vectors, we obtain from Theorem 12.18 that the total Shapley value of a convex multi-choice game, which is the scheme $[\Phi_{ij}^t]_{i \in N, j \in M_i^t}^{t \in \mathcal{M}_+^N}$ with the Shapley value of the multi-choice subgame v_t in each row t, is a limas (see Example 4.2 in [28]) for v.

12.2.3 The Constrained Egalitarian Solution

Now, we introduce (cf. [23]) the multi-choice counterpart of the constrained egalitarian solution of a game $v \in CG^N$ by using an adjusted version of the Dutta-Ray algorithm (cf. [46]). The next proposition will play a key role since it shows that for each convex multi-choice game there exists a unique multi-choice coalition with the largest cumulate number of levels of players among all coalitions with the highest (per one-unit level increase) average worth.

Proposition 12.19. *Let $v \in CMC^{N,m}$ and let*

$$A(v) := \left\{ t \in \mathcal{M}_+^N \mid \alpha(t, v) = \max_{s \in \mathcal{M}_+^N} \alpha(s, v) \right\}.$$

The set $A(v)$ is closed with respect to the join operator $\vee$ and there exists a unique element in $\arg\max_{s \in \mathcal{M}_+^N} \alpha(s, v)$ with the maximal cumulate number of levels.

Proof. Let $\bar{\alpha} = \max_{s \in \mathcal{M}_+^N} \alpha(s, v)$ and take $t^1, t^2 \in A(v)$. We have to prove that $t^1 \vee t^2 \in A(v)$, that is $\alpha(t^1 \vee t^2, v) = \bar{\alpha}$. Since $v(t^1) = \bar{\alpha} \|t^1\|_1$ and $v(t^2) = \bar{\alpha} \|t^2\|_1$, we obtain

$$\begin{aligned}\bar{\alpha}\|t^1\|_1 + \bar{\alpha}\|t^2\|_1 &= v(t^1) + v(t^2) \\ &\leq v(t^1 \vee t^2) + v(t^1 \wedge t^2) \\ &\leq \bar{\alpha}\|t^1 \vee t^2\|_1 + \bar{\alpha}\|t^1 \wedge t^2\|_1 \\ &= \bar{\alpha}\|t^1\|_1 + \bar{\alpha}\|t^2\|_1,\end{aligned}$$

where the first inequality follows from the convexity of v, and the second inequality follows from the definition of $\bar{\alpha}$ and the fact that $v(0) = 0$ (in case $t^1 \wedge t^2 = 0$). This implies that $v(t^1 \vee t^2) = \bar{\alpha}\|t^1 \vee t^2\|_1$. Hence, $t^1 \vee t^2 \in A(v)$, in case $t^1 \wedge t^2 \in A(v)$ as well as in case $\|t^1 \wedge t^2\|_1 = 0$. We can conclude that for any $t^1, t^2 \in A(v)$ not only $t^1 \vee t^2 \in A(v)$ holds true, but also $t^1 \wedge t^2 \in A(v)$ if $t^1 \wedge t^2 \neq 0$. Further, $A(v)$ is closed with respect to finite "unions", where $t^1 \vee t^2$ is seen as the "union" of t^1 and t^2. Now, we note that the set $A(v) \cup \{0\}$ has a lattice structure and $\bigvee_{t \in A(v)} t$ is the largest element in $A(v)$.

Now, we are able to formulate the Dutta-Ray algorithm for convex multi-choice games.

Step 1: Consider $m^1 := m$, $v_1 := v$. Select the unique element in $\arg\max_{s \in \mathcal{M}^N_{m^1} \setminus \{0\}} \alpha(s, v_1)$ with the maximal cumulate number of levels, say s^1. Define $d_{ij} := \alpha(s^1, v_1)$ for each $i \in car(s^1)$ and $j \in M_i^{s^1}$. If $s^1 = m$, then stop; otherwise, go on.

Step p: Suppose that $s^1, s^2, ..., s^{p-1}$ have been defined recursively and $s^1 + s^2 + ... + s^{p-1} \neq m$. Define a new multi-choice game with player set N and maximal participation profile $m^p := m - \sum_{i=1}^{p-1} m^i$. For each multi-choice coalition $s \in \mathcal{M}^N_{m^p}$, define $v_p(s) := v_{p-1}(s + s^{p-1}) - v_{p-1}(s^{p-1})$. The game $v_p \in MC^{N,m^p}$ is convex. Denote by s^p the (unique) largest element in $\arg\max_{s \in \mathcal{M}^N_{m^p} \setminus \{0\}} \alpha(s, v_p)$ and define $d_{ij} := \alpha(s^p, v_p)$ for all $i \in car(s^p)$ and $j \in \left\{\sum_{k=1}^{p-1} s_i^k + 1, \ldots, \sum_{k=1}^{p} s_i^k\right\}$.

In a finite number of steps, say P, where $P \leq |M^+|$, the algorithm will end, and the constructed (level) payoff vector $(d_{ij})_{(i,j) \in M^+}$ is called *the (Dutta-Ray) constrained egalitarian solution* $d(v)$ of the convex multi-choice game v.

The next example illustrates that the constrained egalitarian solution of a convex multi-choice game does not necessarily belong to the imputation set of the game.

Example 12.20. Consider the multi-choice game $v \in CMC^{\{1,2\},m}$ with $m = (3,2)$, $v(0,0) = 0$, $v(1,0) = v(0,1) = 1$, $v(2,0) = v(1,1) = v(0,2) = 2$, $v(2,1) = v(1,2) = 3$, $v(3,0) = v(2,2) = 5$, $v(3,1) = 6$, $v(3,2) = 12$. The constrained egalitarian allocation is $d(v) =$

$(2.4, 2.4, 2.4, 2.4, 2.4)$. Note that $d_{13} = 2.4 < v(3e^1) - v(2e^1) = 5 - 2 = 3$. Hence, $d(v) \notin I(v)$.

Remark 12.21. Note that the above described Dutta-Ray algorithm determines in P steps for each $v \in CMC^{N,m}$ a unique sequence of (per one-unit level increase) average values $\alpha_1, \alpha_2, \ldots, \alpha_P$ with $\alpha_p := \alpha(s^p, v_p)$ for each $p \in \{1, \ldots, P\}$, and a unique sequence of multi-choice coalitions in $\mathcal{M}_+^N$, which we denote by $t^1 := s^1$, $t^2 := s^1 + s^2$,..., $t^p := s^1 + \ldots + s^p$,..., $t^P := s^1 + \ldots + s^P = m$. Thus, a unique path $\langle t^0, t^1, \ldots, t^P \rangle$, with $t^0 = 0$ from 0 to m is obtained, to which we can associate a suitable ordered partition $D^1, D^2, \ldots, D^P$ of M, such that for all $p \in \{1, \ldots, P\}$,

$$D^p := \left\{ (i,j) \mid i \in car(t^p - t^{p-1}),\ j \in \{t_i^{p-1} + 1, \ldots, t_i^p\} \right\},$$

where for each $(i,j) \in D^p$, $d_{ij} = \alpha_p$, and the coalition $t^p - t^{p-1}$ is the maximal participation profile in the "box" D^p with average worth α_p. Note that each other participation profile in D^p can be expressed as $s \wedge t^p - s \wedge t^{p-1} + t^{p-1}$, where $s \in \mathcal{M}_+^N$. Clearly, the average worth of such a participation profile is weakly smaller than α_p, and Proposition 11.28 holds true, too. Hence, $\alpha_p \geq \alpha_{p+1}$ for all $p \in \{1, ..., P-1\}$.

Next, we prove in Theorem 12.23 that the constrained egalitarian solution for convex multi-choice games has similar properties as the constrained egalitarian solution for traditional convex games. We need the following lemma.

Lemma 12.22. *Let $v \in CMC^{N,m}$. Let P be the number of steps in the Dutta-Ray algorithm for constructing the constrained egalitarian solution $d(v)$ of v, and let $t^1, t^2, \ldots, t^P$ be the corresponding sequence of multi-choice coalitions in $\mathcal{M}_+^N$. Then, for each $s \in \mathcal{M}_+^N$ and each $p \in \{1, ..., P\}$,*

$$v(s \wedge t^p - s \wedge t^{p-1} + t^{p-1}) - v(t^{p-1}) \geq v(s \wedge t^p) - v(s \wedge t^{p-1}).$$

Proof. First, notice that, for each $i \in N$,

$$\min\{s_i, t_i^{p-1}\} = \min\{\min\{s_i, t_i^p\}, t_i^{p-1}\}$$

because $t_i^p \geq t_i^{p-1}$, implying that $s \wedge t^{p-1} = (s \wedge t^p) \wedge t^{p-1}$.

Second, notice that, for $i \in N$, either $\min\{s_i, t_i^{p-1}\} = t_i^{p-1}$ or $\min\{s_i, t_i^{p-1}\} = s_i$, and in both situations we have

$$\min\{s_i, t_i^p\} - \min\{s_i, t_i^{p-1}\} + t_i^{p-1} = \max\{\min\{s_i, t_i^p\}, t_i^{p-1}\},$$

implying that

$$(s \wedge t^p) - (s \wedge t^{p-1}) + t^{p-1} = (s \wedge t^p) \vee t^{p-1}.$$

Now, by convexity of v (with $s \wedge t^p$ in the role of s and t^{p-1} in the role of t), we obtain

$$v((s \wedge t^p) \vee t^{p-1}) + v(s \wedge t^{p-1}) \geq v(s \wedge t^p) + v(t^{p-1}).$$

Theorem 12.23. *Let $v \in CMC^{N,m}$. Then the constrained egalitarian allocation $d(v) = (d_{ij})_{i \in N, j \in M_i^+}$ satisfies the following properties:*

(i) $d(v) \in \mathcal{PC}(v)$;
(ii) $d(v)$ Lorenz dominates each $x \in \mathcal{PC}(v)$;
(iii) $d(v) \in EDC(v)$.

Proof. (i) Let P be the number of steps in Dutta-Ray algorithm, $t^1, t^2, \ldots, t^P$ be the corresponding sequence of multi-choice coalitions in $\mathcal{M}_+^N$, and $\alpha_1, \alpha_2, \ldots, \alpha_P$ be the sequence of average values of these coalitions. Note that each $s \in \mathcal{M}_+^N$ can be expressed as

$$s = (s \wedge t^1) + (s \wedge t^2 - s \wedge t^1) + \ldots + (s \wedge t^P - s \wedge t^{P-1}),$$

where some of the terms could be zero. Then, by definition of $D(s)$ and α_p, $p \in \{1, \ldots, P\}$, $D(s)$ can be rewritten as follows:

$$\begin{aligned}
D(s) &= \sum_{i \in N} \sum_{j=1}^{s_i} d_{ij} \\
&= \|s \wedge t^1\|_1 \alpha_1 + \|s \wedge t^2 - s \wedge t^1\|_1 \alpha_2 + \ldots + \|s \wedge t^P - s \wedge t^{P-1}\|_1 \alpha_P \\
&= \|s \wedge t^1\|_1 \frac{v(t^1)}{\|t^1\|_1} + \|s \wedge t^2 - s \wedge t^1\|_1 \frac{v(t^2) - v(t^1)}{\|t^2 - t^1\|_1} \\
&\quad + \ldots + \|s \wedge t^P - s \wedge t^{P-1}\|_1 \frac{v(t^P) - v(t^{P-1})}{\|t^P - t^{P-1}\|_1}.
\end{aligned}$$

Now, in view of Remark 12.21, we obtain

$$
\begin{aligned}
D(s) &\geq \|s \wedge t^1\|_1 \frac{v(t^1)}{\|s \wedge t^1\|_1} \\
&\quad + \|s \wedge t^2 - s \wedge t^1\|_1 \frac{v((s \wedge t^2) - (s \wedge t^1) + t^1) - v(t^1)}{\|s \wedge t^2 - s \wedge t^1\|_1} + \dots \\
&\quad + \|s \wedge t^P - s \wedge t^{P-1}\|_1 \frac{v((s \wedge t^P) - (s \wedge t^{P-1}) - t^{P-1}) - v(t^{P-1})}{\|s \wedge t^P - s \wedge t^{P-1}\|_1} \\
&\geq \|s \wedge t^1\|_1 \frac{v(s \wedge t^1)}{\|s \wedge t^1\|_1} + \|s \wedge t^2 - s \wedge t^1\|_1 \frac{v(s \wedge t^2) - v(s \wedge t^1)}{\|s \wedge t^2 - s \wedge t^1\|} + \dots \\
&\quad + \|s \wedge t^P - s \wedge t^{P-1}\|_1 \frac{v(s \wedge t^P) - v(s \wedge t^{P-1})}{\|s \wedge t^P - s \wedge t^{P-1}\|_1} \\
&= v(s \wedge t^1) + (v(s \wedge t^2) - v(s \wedge t^1)) + \dots \\
&\quad + (v(s \wedge t^P) - v(s \wedge t^{P-1})) \\
&= v(s \wedge t^P) = v(s),
\end{aligned}
$$

where the last inequality follows from Lemma 12.22. Hence, $D(s) \geq v(s)$ for each $s \in \mathcal{M}_+^N$. Finally, $D(m) = v(m)$ follows from the constructive definition of d, too.

(ii) Let $x \in \mathcal{PC}(v)$. We notice that the Lorenz domination relation (cf. Section 4.1) can be directly extended to level payoff vectors and prove by backward induction that $x \succ_L d$ implies $x = d$. Suppose that $0 = t^0, t^1, \dots, t^P = m$ is the sequence of the Dutta-Ray multi-choice coalitions in $\mathcal{M}^N$ for d (see Remark 11.26). First, we prove the induction basis, i.e.,

$$
\begin{aligned}
d_{ij} = x_{ij} &\text{ for each } (i,j) \in (t^{P-1}, t^P], \\
&\text{i.e., for all } j \text{ s.t. } t_i^{P-1} < j \leq t_i^P, i \in N. \qquad (12.13)
\end{aligned}
$$

By Proposition 11.28 (see Remark 12.21), the smallest elements of $d = (d_{ij})_{i \in N, j \in M_i^+}$ correspond precisely to elements $(i,j) \in (t^{P-1}, t^P]$, and there

$$
d_{ij} = \frac{v(t^P) - v(t^{P-1})}{\|t^P - t^{P-1}\|_1} = \alpha_P.
$$

Since $x \succ_L d$, it follows that $x_{ij} \geq d_{ij}$ for all $(i,j) \in (t^{P-1}, t^P]$, implying that

$$
\begin{aligned}
x((t^{P-1}, t^P]) &= \sum_{(i,j) \in (t^{P-1}, t^P]} x_{ij} \\
&\geq \sum_{(i,j) \in (t^{P-1}, t^P]} d_{ij} = d((t^{P-1}, t^P]) \\
&= \alpha_P \|t^P - t^{P-1}\|_1.
\end{aligned}
$$

Suppose that (12.13) does not hold. Then, we obtain

$$x((t^{P-1},t^P]) > d(t^{P-1},t^P] = v(t^P) - v(t^{P-1}).$$

But, since $x \in \mathcal{PC}(v)$, we also have

$$\begin{aligned} x((t^{P-1},t^P]) &= x((0,t^P]) - x((0,t^{P-1}]) \\ &= v(t^P) - x((0,t^{P-1}]) \\ &\leq v(t^P) - v(t^{P-1}) \\ &= \alpha_P \|t^P - t^{P-1}\|_1, \end{aligned}$$

where the second equality follows from the efficiency condition for precore elements and the inequality from the stability conditions for precore elements. So, we conclude that (12.13) holds true.

Now, we prove the induction step, i.e., for each $k \in \{P-1,\ldots,1\}$ it holds true that

$$\begin{aligned} &\text{if } d_{ij} = x_{ij} \text{ for each } (i,j) \in (t^k,t^P], \text{ then} \\ &\quad d_{ij} = x_{ij} \text{ for each } (i,j) \in (t^{k-1},t^P]. \end{aligned} \tag{12.14}$$

Suppose $d = x$ on $(t^k,t^P]$. Then, by Proposition 11.28, the worst $\|t^P - t^k\|_1$ elements of d and x are in $(t^k,t^P]$ and "coincide". Since $x \succ_L d$ we have:

(*) $x_{ij} \geq \alpha_k = d_{ij}$ for all $(i,j) \in (t^{k-1},t^k]$, $k \in \{1,...,P\}$, and

(**) $x((0,t^k]) = \sum_{(i,j)\in(0,t^k]} x_{ij} = v(t^k)$ because $v(t^P) - v(t^k) = d((t^k,t^P]) = x((t^k,t^P])$ and $v(t^P) = x\left((0,t^P]\right)$.

Then, $x \in \mathcal{PC}(v)$ implies

(***) $x((t^{k-1},t^k]) = x((0,t^k]) - x((0,t^{k-1}]) \leq v(t^k) - v(t^{k-1}) = \alpha_k\|t^k - t^{k-1}\|_1$,

where the inequality follows from (**) and the stability conditions. Then, from (*) and (***) it follows that $x_{ij} = d_{ij}$ for $(i,j) \in (t^{k-1},t^k]$, and so, $x = d$ on $(t^{k-1},t^P]$.

(iii) According to (i) and Theorem 11.14, we have $d(v) \in \mathcal{PC}(v) \subset EDC(v)$.

The next theorem provides an axiomatic characterization of the constrained egalitarian solution on the class of convex multi-choice games in line with Theorem 3.3 in [65], using the following multi-choice versions of the properties of efficiency, equal division stability and max-consistency.

Given a single-valued solution $\Lambda : CMC^{N,m} \to \mathbb{R}^{\sum_{i\in N} m_i}$, we denote by s^m the multi-choice coalition such that for each $i \in N$, for all $k \in \{1,\ldots,s_i^m\}$, $\Lambda_{ik}(v) = \max_{(i,j)\in M^+} \Lambda_{ij}(v)$. We say that Λ satisfies

- *Efficiency*, if for all $v \in CMC^{N,m}$: $\sum_{i \in N} \sum_{j=1}^{m_i} \Lambda_{ij}(v) = v(m)$;
- *Equal division stability*, if for all $v \in CMC^{N,m}$: $\Lambda(v) \in EDC(v)$;
- *Max-consistency*, if for all $v \in CMC^{N,m}$ and all $(i,j) \in M^+$, $\Lambda_{ij}(v) = \Lambda_{ij}(v^{-s^m})$, where v^{-s^m} is the multi-choice game defined by $v^{-s^m}(t) = v(t + s^m) - v(s^m)$ for all $t \in \mathcal{M}^N_{m-s^m}$.

Theorem 12.24. *There is a unique solution on $CMC^{N,m}$ satisfying the properties efficiency, equal division stability and max consistency, and it is the constrained egalitarian solution.*

12.2.4 Properties of Solution Concepts

As in the case of convex crisp games, solution concepts on convex multi-choice games have nice properties.

First, note that by Theorem 12.11 the Shapley value $\Phi(v)$ (cf. Definition 11.18) of $v \in CMC^{N,m}$ belongs to the core $C(v)$ of v. Moreover, we have already seen that the corresponding extended Shapley value is a limas of the game and hence, the convexity of a multi-choice game is a sufficient condition for the existence of such monotonic allocation schemes.

Second, the core of a convex multi-choice game is the unique stable set of the game.

Theorem 12.25. *Let $v \in CMC^{N,m}$. Then $C(v)$ is the unique stable set.*

Proof. Using Corollary 12.16 we see that $C(v) \neq \emptyset$. Hence, it follows from Theorem 11.9 that $C(v) = DC(v)$. So, by Theorem 11.12(ii) we know that it suffices to prove that $C(v)$ is a stable set.

Internal stability of $C(v)$ is obvious. To show external stability, let $x \in I(v) \setminus C(v)$. We construct $z \in C(v)$ that dominates x. First, we choose $s \in \mathcal{M}^N_+$ such that

$$|car(s)|^{-1} (v(s) - X(s)) = \max_{t \in \mathcal{M}^N_+} |car(t)|^{-1} (v(t) - X(t)).$$

Since $x \notin C(v)$ it holds that

$$\left|car(s)^{-1}\right| (v(s) - X(s)) > 0. \tag{12.15}$$

Now, let σ be an order that is admissible for v with the property that there exists k such that $s = s^{\sigma,k}$. Then (cf. Theorem 12.11) the corresponding marginal vector w^σ is an element of $C(v)$ and, moreover, it holds that $\sum_{i \in N} \sum_{j=1}^{s_i} w^\sigma_{ij} = v(s)$. For notational convenience we set

$y := w^\sigma$. We define the payoff vector z by $z_{ij} = x_{ij}$ if $i \in car(s)$ and $2 \le j \le s_i$; $z_{ij} = x_{i1} + |car(s)|^{-1}(v(s) - X(s))$ if $i \in car(s)$ and $j = 1$; $z_{ij} = y_{ij}$ if $i \notin car(s)$ or $i \in car(s)$ and $j > s_i$; $z_{i0} = 0$.

Using the fact that $x, y \in I(v)$ and recalling (12.15), it can be easily seen that z is level increase rational. Further, $Z(m) = X(s) + (v(s) - X(s)) + (Y(m) - Y(s)) = v(s) + (v(m) - v(s)) = v(m)$, where the second equality follows from the way we choose y. This shows that z is also efficient and, hence, $z \in I(v)$. Since $Z(s) = v(s)$ and $Z_{is_i} = X_{is_i} + |car(s)|^{-1}(v(s) - X(s)) > X_{is_i}$ for all $i \in car(s)$, it holds that $z \,\mathrm{dom}_s\, x$.

The only part that is left to prove is $z \in C(v)$. So, let $t \in \mathcal{M}_+^N$. We distinguish two cases.

(a) If $car(t) \cap car(s) = \emptyset$, then $Z(t) = Y(t) \ge v(t)$ since $y \in C(v)$.

(b) If $car(t) \cap car(s) \ne \emptyset$, then

$$\begin{aligned} Z(s \wedge t) &= (Z - X)(s \wedge t) + X(s \wedge t) \\ &= |car(s) \cap car(t)| \,.\, |car(s)|^{-1}(v(s) - X(s)) + X(s \wedge t) \\ &\ge v(s \wedge t) - X(s \wedge t) + X(s \wedge t) = v(s \wedge t), \end{aligned} \tag{12.16}$$

where the inequality follows from (12.15). Hence,

$$\begin{aligned} Z(t) &= \sum_{i \in N} \sum_{j=1}^{s_i \wedge t_i} z_{ij} + \sum_{i \in N: t_i > s_i} \sum_{j=s_i+1}^{t_i} y_{ij} \\ &= Z(s \wedge t) + \sum_{i \in N} \sum_{j=1}^{s_i \vee t_i} y_{ij} - \sum_{i \in N: s_i \ge t_i} \sum_{j=1}^{s_i} y_{ij} - \sum_{i \in N: s_i < t_i} \sum_{j=1}^{s_i} y_{ij} \\ &= Z(s \wedge t) + Y(s \vee t) - Y(s) \\ &\ge v(s \wedge t) + v(s \vee t) - v(s), \end{aligned} \tag{12.17}$$

where the last equality follows from (12.16) and the fact that $y \in C(v)$ is such that $Y(s) = v(s)$. Using convexity of v, we see that the last expression in (12.17) is larger or equal of $v(t)$. This completes the proof of the theorem.

Finally, in a straightforward way, but technically cumbersome, we can extend the properties of the equal split-off set for convex crisp games (cf. Subsection 5.2.4) to convex multi-choice games. Specifically, for each game $v \in CMC^{N,m}$, the equal split-off set $ESOS(v)$ consists of a single element which is the constrained egalitarian solution $d(v)$ of the game v. The proof of this result follows the lines of Lemma 5.27 and Theorem 5.28.

12.3 Multi-Choice Clan Games

12.3.1 Basic Characterizations

In this section we introduce a new class of multi-choice games, which we call multi-choice clan games. As in the traditional model of clan games (cf. Section 5.3), the set of players consists of two disjoint groups, a fixed (powerful) clan with "yes-or-no" choices and a group of (non-powerful) non-clan members, but in the multi-choice clan game non-clan members may have more options for cooperation. Specifically, each non-clan member can participate at any level in a given finite set, whereas each clan member can be either active or abstain from cooperation. However, the active status (i.e. participation level 1) for all clan members is a necessary condition for generating a positive reward for any coalition containing the clan and at least one non-clan member.

Let $N = (N \setminus C, C)$ be the set of players, where C stands for the clan and $N \setminus C$ for the group of non-clan members. For further use, we denote by $\mathcal{M}^{N,C}$ the set of multi-choice coalitions with player set N and fixed clan C. For each $s \in \mathcal{M}^{N,C}$ we denote its restrictions to $N \setminus C$ and C, by $s_{N\setminus C}$ and s_C, respectively. Note that, in a multi-choice clan game, the maximal participation profile is of the form $m = (m_{N\setminus C}, 1_C)$.

We also denote by $\mathcal{M}^{N,1_C}$ the set of multi-choice coalitions $s \in \mathcal{M}^{N,C}$ with $s_C = 1_C$, that is coalitions in which all clan members fully participate. We also use the notation $\mathcal{M}_+^{N,1_C} = \mathcal{M}^{N,1_C} \setminus \{0\}$. Multi-choice clan games are defined here by using the veto power of clan members, the monotonicity property of the characteristic function and a (level) union property regarding non-clan members' participation in multi-choice coalitions containing at least all clan members at participation level 1.

Definition 12.26. *A game $\langle N, (m_{N\setminus C}, 1_C), v\rangle$ is a* ***multi-choice clan game*** *if the characteristic function $v : \mathcal{M}^{N,C} \to \mathbb{R}$ satisfies*

(i) Clan property: $v(s) = 0$ if $s_C \neq 1_C$;
(ii) Monotonicity property: $v(s) \leq v(t)$ for all $s, t \in \mathcal{M}^{N,C}$ with $s \leq t$;
(iii) (Level) Union property: For each $s \in \mathcal{M}^{N,1_C}$,

$$v(m) - v(s) \geq \sum_{i \in N\setminus C} w_{is_i^+}(m, v),$$

where, for each $i \in N \setminus C$, $w_{is_i^+}(m, v)$ denotes the marginal contribution of the bundle of those levels of player i which are higher than the participation level of i in s.

In the sequel, we simply use v instead of $\langle N, (m_{N\setminus C}, 1_C), v\rangle$. For further use, we denote by $MC_C^{N,m}$ the set of multi-choice games with a fixed non-empty and finite set of players N, fixed non-empty clan C, and maximal participation profile m. We notice that $MC_C^{N,m}$ is a convex cone in $MC^{N,m}$, that is for all $v, w \in MC_C^{N,m}$ and for all $p, q \in \mathbb{R}_+$, $pv+qw \in MC_C^{N,m}$, where $\mathbb{R}_+$ denotes the set of non-negative real numbers.

The next theorem gives an explicit description of the core of a multi-choice clan game.

Theorem 12.27. *Let* $v \in MC_C^{N,m}$. *Then,*

$$C(v) = \{x : M \to \mathbb{R}_+ \mid X(m) = v(m);$$
$$\sum_{k=j}^{m_i} x_{ik} \leq v(m) - v(m_{-i}, j-1), \forall i \in N \setminus C,\ j \in M_i^+\}.$$

Proof. We denote by $B(v)$ the set in the right-hand side of the above equality. We prove first that $C(v) \subset B(v)$. Let $v \in MC_C^{N,m}$ and let $x \in C(v)$. Note that non-negativity of x follows from the clan property and the monotonicity property; so, $x_{ij} \geq 0$ for all $i \in N$ and $j \in M_i^+$. Clearly, the efficiency condition holds true. To prove the upper boundness of the cumulate payoffs for each bundle of highest levels of each non-clan member, we note first that the payoff of each multi-choice coalition $\tilde{s} = (m_{-i}, j-1) \in \mathcal{M}^{N,1_C}$, $i \in N \setminus C$, $j \in \{1, \ldots, m_i\}$, can be expressed as

$$X(\tilde{s}) = X(m) - \sum_{k=j}^{m_i} x_{ik} = v(m) - \sum_{k=j}^{m_i} x_{ik},$$

where the second equality follows from the efficiency of x.

Now, the stability condition $X(\tilde{s}) \geq v(\tilde{s})$ implies that

$$v(m) - \sum_{k=j}^{m_i} x_{ik} \geq v(m_{-i}, j-1),$$

or, equivalently,

$$\sum_{k=j}^{m_i} x_{ik} \leq v(m) - v(m_{-i}, j-1).$$

Hence, $x \in B(v)$.

Now, we prove the converse inclusion. Let $x \in B(v)$. The level-increase rationality of x follows from the clan property, since $x_{ij} \geq 0 = v(0_{-i}, j) - v(0_{-i}, j-1)$ for all $i \in N$ and $j \in M_i^+$. We only need to prove that $X(s) \geq v(s)$ for each $s \in \mathcal{M}^{N,C}$.

Clearly, for each $s \in \mathcal{M}^{N,C}$ with $s_C \neq 1_C$, we have, by the clan property, $X(s) = \sum_{i \in N} \sum_{j=1}^{s_i} x_{ij} \geq 0 = v(s)$.

We focus now on stability conditions for multi-choice coalitions $s \in \mathcal{M}^{N,C}$ with $s_C = 1_C$.

By the (level) union property, we obtain

$$v(m) - v(s) \geq \sum_{i \in N \setminus C} w_{is_i^+}(m, v).$$

Further,

$$X(s) + \sum_{i \in N \setminus C} \sum_{k=s_i+1}^{m_i} x_{ik} - v(s) = X(m) - v(s) = v(m) - v(s)$$

$$\geq \sum_{i \in N \setminus C} w_{is_i^+}(m, v),$$

implying that

$$X(s) \geq v(s) + \sum_{i \in N \setminus C} (w_{is_i^+}(m, v) - \sum_{k=s_i+1}^{m_i} x_{ik}).$$

Note that $w_{is_i^+}(m, v) = v(m) - v(m_{-i}, s_i) \geq \sum_{k=s_i+1}^{m_i} x_{ik}$, for each $i \in N \setminus C$, since $x \in B(v)$. Therefore, we obtain $w_{is_i^+}(m, v) - \sum_{k=s_i+1}^{m_i} x_{ik} \geq 0$ for each $i \in N \setminus C$. Hence, $X(s) \geq v(s)$ for each $s \in \mathcal{M}^{N,1_C}$.

Corollary 12.28. *Let $v \in MC_C^{N,m}$ and let $t \in \mathcal{M}^{N,1_C}$. If the subgame v_t is a clan game, then its core is described by*

$$C(v_t) = \{x : M^t \to \mathbb{R}_+ \mid X(t) = v(t);$$
$$\sum_{k=j}^{t_i} x_{ik} \leq v(t) - v(t_{-i}, j-1), \ \forall i \in car(t_{N \setminus C}), j \in M_i^t\},$$

where $M_i^t = \{1, \ldots, t_i\}$ and $M^t = \{(i, j) \mid i \in car(t), \ j \in M_i^t\}$.

Proof. It follows straightforwardly from Theorem 12.27, by taking into account that $t = (t_{N \setminus C}, 1_C)$ is the "grand coalition" in the subgame v_t.

Multi-choice clan games for which the clan consists of one player are called multi-choice big boss games; we notice that in case $m_{N \setminus C} = 1_{N \setminus C}$ these games are big boss games in the terminology of [27]. The model of a multi-choice clan game where the clan consists of at least two members is an extension of the model of a clan game (cf. [122]).

In the rest of this section, we focus on total clan games with multi-choice coalitions.

Definition 12.29. *A game $v \in MC_C^{N,m}$ is a* ***multi-choice total clan game*** *with clan C if all its subgames v_t, $t \in \mathcal{M}^{N,1_C}$, are clan games with clan C.*

We denote the set of all multi-choice total clan games $v \in MC_C^{N,m}$ by $TMC_C^{N,m}$.

The next theorem provides a characterization of multi-choice total clan games.

Theorem 12.30. *Let $v \in MC^{N,m}$ and $C \in 2^N \setminus \{\emptyset\}$ with $m_C = 1_C$. Then, the following assertions are equivalent:*

(i) $v \in TMC_C^{N,m}$;

(ii) v is monotonic, each player $i \in C$ is a veto player, and for all $s, t \in \mathcal{M}^{N,1_C}$ with $s \leq t$,

$$v(t) - v(s) \geq \sum_{i \in car(t_{N \setminus C})} w_{is_i^+}(t, v); \tag{12.18}$$

(iii) v is monotonic, each player $i \in C$ is a veto player, and for each $i \in car(s_{N \setminus C})$ and for all $s, t \in \mathcal{M}^{N,1_C}$ with $s \leq t$ and $s_i = t_i$,

$$v(t) - v(t - e^i) \leq v(s) - v(s - e^i). \tag{12.19}$$

Proof. (i)$\leftrightarrow$(ii): Relation (12.18) simply writes out the (level) union property of multi-choice subgames. In the sequel, we refer to relation (12.18) as *the total (level) union property* of v.

(ii)$\rightarrow$(iii): It suffices to prove that (12.18)$\rightarrow$(12.19). We note that inequality (12.19) expresses a *total concavity property* of v which reflects the fact that the same one-unit level decrease of a non-clan member in coalitions containing at least all clan members at participation level 1 and where that non-clan member has the same participation level, could be more beneficial in smaller such coalitions than in larger ones.

Let $s \in \mathcal{M}^{N,1_C}$ and let $i \in car(s_{N\setminus C})$. Consider the coalition $s + e^k$ obtained from s when one the other non-clan members, say k, increases his participation with one unit. We prove first

$$w_{is_i}(s,v) \geq w_{is_i}(s+e^k,v). \tag{12.20}$$

$$\begin{aligned} & w_{is_i}(s,v) + w_{k,s_k+1}(s+e^k,v) \\ &= (v(s) - v(s_{-i}, s_i - 1)) + (v(s+e^k) - v(s_{-k}, s_k)) \\ &= v(s+e^k) - v(s_{-i}, s_i - 1). \end{aligned} \tag{12.21}$$

The total union property (with $s + e^k$ in the role of t and $(s_{-i}, s_i - 1)$ in the role of s) yields

$$v(s+e^k) - v(s_{-i}, s_i - 1) \geq w_{is_i}(s+e^k,v) + w_{k,s_k+1}(s+e^k,v). \tag{12.22}$$

From (12.21) and (12.22) we conclude that (12.20) holds true. Denote by $\{i_1, \ldots, i_q\}$ the set of levels which are involved in t but not in s. Repeated application of (12.20) yields

$$w_{is_i}(s,v) \geq w_{is_i}(s+e^{i_1},v) \geq \ldots \geq w_{is_i}(s+(e^{i_1}+\ldots+e^{i_q})) = w_{is_i}(t,v).$$

Hence, the total concavity property (12.19) holds true.

(iii)→(ii): We simply prove that (12.19)→(12.18). Let $s, t \in \mathcal{M}^{N,1_C}$ with $s \leq t$. Denote by $\{i_1, \ldots, i_q\}$ the set of levels which are involved in t but not in s.

Then,

$$\begin{aligned} v(t) - v(s) &= \sum_{i \in car(t_{N\setminus C})} \sum_{k=1}^{q} w_{i,i_k}(s + (e^{i_1} + \ldots + e^{i_k}), v) \\ &\geq \sum_{i \in car(t_{N\setminus C})} \sum_{k=1}^{q} w_{i,i_k}(t,v) = \sum_{i \in car(t_{N\setminus C})} w_{is_i^+}(t,v). \end{aligned}$$

Hence, the total union property (12.18) holds true.

12.3.2 Bi-Monotonic Allocation Schemes

The notion of a level-increase monotonic allocation scheme (limas) for a multi-choice game was recently introduced in [28] where it was also shown that convexity of the game is a sufficient condition for the existence of a limas. Inspired by [27] and [122] who study the existence of bi-monotonic allocation schemes for a total big boss and clan games, we focus now on bi-(level-increase) monotonic allocation schemes (bi-limas) for multi-choice total clan games.

Definition 12.31. *Let $v \in TMC_C^{N,m}$ and let $t \in \mathcal{M}^{N,1_C}$. A scheme, $[a_{ij}^t]_{i\in N, j\in M_i^t}^{t\in\mathcal{M}^{N,1_C}}$ where $M_i^t = \{1,\dots,t_i\}$, is called a* ***bi-(level-increase) monotonic allocation scheme*** *(bi-limas) if the following two conditions hold:*

(i) stability: $a^t \in C(v_t)$ for all $t \in \mathcal{M}^{N,1_C}$;
(ii) bi-monotonicity w.r.t. one-unit level increase: For all $s,t \in \mathcal{M}^{N,1_C}$ with $s \le t$, we have:
(ii.1) $a_{i1}^s \le a_{i1}^t$ for each $i \in C$;
(ii.2) $a_{ij}^s \ge a_{ij}^t$ for each $i \in car(s_{N\setminus C})$ and each $j \in \{1,\dots,s_i\}$.

We study this kind of bi-monotonic allocation schemes by means of suitably defined compensation-sharing rules on the class of multi-choice (total) clan games.

Definition 12.32. *Let $N \setminus C = \{1,\dots,q\}$ and $C = \{q+1,\dots,n\}$. For each $\alpha \in [0,1]^q$ and $\beta \in \Delta(C) = \Delta(\{q+1,\dots,n\}) = \{z_+^{n-q} \mid \sum_{i=q+1}^{n} z_i = 1\}$, the* ***compensation-sharing rule*** *based on α and β, $\psi^{\alpha,\beta} : MC_C^{N,m} \to \mathbb{R}^{M^+}$, is defined by*

$$\psi_{ij}^{\alpha,\beta}(v) = \begin{cases} \alpha_i(v(m) - v(m_{-i},0)) & , i \in N \setminus C,\ j = 1; \\ 0 & , i \in N \setminus C, \\ & \quad j \in M_i^+ \setminus \{1\}; \\ \beta_i[v(m) - \sum_{j\in N\setminus C} \alpha_j(v(m) - v(m_{-j},0))] & , i \in C,\ j = 1 \end{cases}$$

for each $i \in N$ and $j \in M_i^+$.

The i-th coordinate of the compensation vector α indicates that level 1 of non-clan member i gets as payoff, in view of its decisive role for multi-choice cooperation, the part $\alpha_i w_i(m,v)$ of the marginal contribution of this player to the grand coalition m. Then, the remainder, $v(m) - \sum_{j\in N\setminus C} \alpha_j w_j(m,v)$, is distributed over the clan members. For each clan member i, the i-th coordinate β_i of the sharing vector β determines the share $\beta_i[v(m) - \sum_{j\in N\setminus C} \alpha_j w_j(m,v)]$.

Theorem 12.33. *Let $MC_C^{N,m}$ be the cone of multi-choice clan games with clan C. Then,*
(i) $\psi^{\alpha,\beta}$ is additive for each $\alpha \in [0,1]^{N\setminus C}$ and each $\beta \in \Delta(C)$;
(ii) $\psi^{\alpha,\beta}$ is stable, that is $\psi^{\alpha,\beta}(v) \in C(v)$ for each $v \in MC_C^{N,m}$.

Proof. (i) Let $\alpha \in [0,1]^{N\setminus C}$ and $\beta \in \Delta(C)$. For all $v, w \in MC_C^{N,m}$ and all $p, q \in \mathbb{R}_+$, it holds

$$\psi^{\alpha,\beta}(pv + qw) = p\psi^{\alpha,\beta}(v) + q\psi^{\alpha,\beta}(w).$$

Hence, $\psi^{\alpha,\beta}$ is additive on the cone of multi-choice clan games.

(ii) Let $v \in MC_C^{N,m}$. From Theorem 12.27 and $\sum_{i\in C} \beta_i = 1$ we obtain that $\psi^{\alpha,\beta}(v) \in C(v)$.

In Theorem 12.36 we prove that the family of compensation-sharing rules $\psi^{\alpha,\beta}$ plays a key role for the existence of bi-limas for a subclass of multi-choice total clan games. However, we need to establish some preliminary results.

Lemma 12.34. *Let $v \in TMC_C^{N,m}$ and $s, t \in \mathcal{M}^{N,1_C}$ with $s \leq t$. Then,*

$$v(t) - v(s) \geq \sum_{i \in car((t-s)_{N\setminus C})} w_i(t, v).$$

Proof. Note that $s \leq t$ implies $car(s_{N\setminus C}) \subset car(t_{N\setminus C})$. We denote $car(t_{N\setminus C}) - car(s_{N\setminus C})$ by $car((t-s)_{N\setminus C})$. From the total (level) union property we obtain

$$\begin{aligned}
v(t) - v(s) &\geq \sum_{i \in car(t_{N\setminus C})} w_{is_i^+}(t, v) \\
&= \sum_{i \in car(s_{N\setminus C})} w_{is_i^+}(t, v) + \sum_{i \in car((t-s)_{N\setminus C})} w_{i0^+}(t, v) \\
&= \sum_{i \in car(s_{N\setminus C})} (v(t) - v(t_{-i}, s_i)) + \sum_{i \in car((t-s)_{N\setminus C})} w_{is_i^+}(t, v) \\
&\geq \sum_{i \in car((t-s)_{N\setminus C})} w_{is_i^+}(t, v),
\end{aligned}$$

where the last inequality holds true because, by monotonicity of v, $v(t) - v(t_{-i}, s_i) \geq 0$ for each $i \in car(s_{N\setminus C})$.

Lemma 12.35. *Let $v \in TMC_C^{N,m}$ and $s, t \in \mathcal{M}^{N,1_C}$ such that $s \leq t$. Then,*

$$v(t) - v(s) - \sum_{i \in car((t-s)_{N\setminus C})} \alpha_i w_i(t, v) \geq 0.$$

Proof. First, by non-negativity of α_i and $w_i(t,v)$, and since $\alpha_i \leq 1$ for each $i \in N \setminus C$ and $t \in \mathcal{M}^{N,1_C}$, we have $\alpha_i w_i(t,v) \leq w_i(t,v)$, implying that

$$-\sum_{i \in car((t-s)_{N\setminus C})} \alpha_i w_i(t,v) \geq -\sum_{i \in car((t-s)_{N\setminus C})} w_i(t,v).$$

Second, by Lemma 12.34 we have

$$v(t) - v(s) - \sum_{i \in car((t-s)_{N\setminus C})} w_i(t,v) \geq 0.$$

Therefore,

$$\begin{aligned} v(t) - v(s) &- \sum_{i \in car((t-s)_{N\setminus C})} \alpha_i w_i(t,v) \\ \geq v(t) - v(s) &- \sum_{i \in car((t-s)_{N\setminus C})} w_i(t,v) \geq 0. \end{aligned}$$

It turns out that for a subclass of multi-choice total clan games compensation-sharing rules $\psi^{\alpha,\beta}$ with $\alpha \in [0,1]^{N\setminus C}$ and $\beta \in \Delta(C)$ generate bi-(level-increase) monotonic allocation schemes.

Theorem 12.36. *Let $v \in TMC_C^{N,m}$ be such that, for each $s,t \in \mathcal{M}^{N,1_C}$ with $s \leq t$ and each $i \in car(s_{N\setminus C})$,*

$$v(t) - v(t_{-i},0) \leq v(s) - v(s_{-i},0). \tag{12.23}$$

Then, for each $\alpha \in [0,1]^{N\setminus C}$ and $\beta \in \Delta(C)$ the compensation-sharing rule $\psi^{\alpha,\beta}$ generates a bi-limas for v, namely

$$[\psi_{ij}^{\alpha,\beta}(v_t)]_{i\in N, j\in M_i^t}^{t\in\mathcal{M}^{N,1_C}}.$$

Proof. Let $\alpha \in [0,1]^{N\setminus C}$ and $\beta \in \Delta(C)$. For each $t \in \mathcal{M}^{N,1_C}$, in the subgame v_t, the α-based compensation (regardless of β) for each non-clan member $i \in car(t_{N\setminus C})$, $\alpha_i w_i(t,v)$, is fully assigned as payoff to level 1 of that player. So, the α-based compensation for each other level of each non-clan member $i \in car(t_{N\setminus C})$ is simply equal to 0. The amount left for the clan, $v(t) - \sum_{j\in car(t_{N\setminus C})} \alpha_j w_j(t,v)$, is shared based on β. For each clan member $i \in C$, the β-based share in the subgame v_t is $\beta_i[v(t) - \sum_{j\in car(t_{N\setminus C})} \alpha_j w_j(t,v)]$, with $\sum_{i\in C} \beta_i = 1$.

Since v_t is a clan game, by Theorem 12.33(ii) and Corollary 12.28, we have $\psi^{\alpha,\beta}(v_t) \in C(v_t)$.

Now, we focus on the bi-monotonicity property. Let $s, t \in \mathcal{M}^{N,1_C}$ with $s \leq t$. First, we prove that, for each player $i \in car(s_{N\setminus C})$ and for each level $j \in \{1, \ldots, s_i\}$ the compensation $\psi_{ij}^{\alpha,\beta}(v_s)$ is weakly better than the compensation $\psi_{ij}^{\alpha,\beta}(v_t)$. Clearly, $\psi_{ik}^{\alpha,\beta}(v_s) = 0 = \psi_{ik}^{\alpha,\beta}(v_t)$ for $k \in \{2, \ldots, s_i\}$. Further, by (12.23) and non-negativity of α_i, we obtain $\psi_{i1}^{\alpha,\beta}(v_s) = \alpha_i w_i(s, v) \geq \alpha_i w_i(t, v) = \psi_{i1}^{\alpha,\beta}(v_s)$ for each $i \in car(s_{N\setminus C})$.

In the sequel, we cope with the monotonicity condition regarding shares of clan members. Denote by $R_\alpha(v_t)$ the α-based remainder for the clan members in the "grand coalition" $(t_{N\setminus C}, 1_C)$ of the multi-choice clan game v_t. We prove first that, for each $s, t \in \mathcal{M}^{N,1_C}$ with $s \leq t$,

$$R_\alpha(v_t) \geq R_\alpha(v_s). \tag{12.24}$$

We have

$$\begin{aligned}
&R_\alpha(v_t) - R_\alpha(v_s) \\
&= \left(v(t) - \sum_{i \in car(t_{N\setminus C})} \alpha_i w_i(t, v) \right) - \left(v(s) - \sum_{i \in car(s_{N\setminus C})} \alpha_i w_i(s, v) \right) \\
&= \left(v(t) - v(s) - \sum_{i \in car((t-s)_{N\setminus C})} \alpha_i w_i(t, v) \right) \\
&\qquad\qquad + \sum_{i \in car(s_{N\setminus C})} \alpha_i (w_i(s, v) - w_i(t, v)) \\
&\geq v(t) - v(s) - \sum_{i \in car((t-s)_{N\setminus C})} \alpha_i w_i(t, v) \geq 0,
\end{aligned}$$

where the first inequality follows from (12.23) and the non-negativity of α_i, and the last inequality follows from Lemma 12.35.

Hence, relation (12.24) holds true. Now, from the non-negativity of β_i and (12.24) we obtain $\psi_{i1}^{\alpha,\beta}(v_t) \geq \psi_{i1}^{\alpha,\beta}(v_s)$, for each $i \in C$.

Remark 12.37. Let $v \in TCM_C^{N,m}$ and let $s, t \in \mathcal{M}^{N,1_C}$ such that $s \leq t$. Then, for each $i \in car(s_{N\setminus C})$ such that $s_i = t_i$, it holds

$$w_i(t, v) \leq w_i(s, v). \tag{12.25}$$

Proof. Let $s, t \in \mathcal{M}^{N,1_C}$ with $s \leq t$, and let $i \in car(s_{N\setminus C})$ such that $s_i = t_i$. Repeated application of the total concavity property yields

$$v(t_{-i}, s_i) - v(t_{-i}, s_i - 1) \leq v(s_{-i}, s_i) - v(s_{-i}, s_i - 1),$$
$$\vdots$$
$$v(t_{-i}, 1) - v(t_{-i}, 0) \leq v(s_{-i}, 1) - v(s_{-i}, 0).$$

By summing these inequalities, we obtain $v(t) - v(t_{-i}, 0) \leq v(s) - v(s_{-i}, 0)$. Hence, (12.25) holds.

For arbitrary multi-choice total clan games bi-limas does not necessarily exist, even if we consider a weaker version of bi-limas where the (level) monotonicity condition regarding the non-clan members is defined by:

For all $s, t \in \mathcal{M}^{N,1_C}$ with $s \leq t$, we have $a^s_{ij} \geq a^t_{ij}$ for each $i \in car(s_{N\setminus C})$ such that $s_i = t_i$ and each $j \in \{1, \dots, s_i\}$.

Then, by Remark 12.37, the (level) monotonicity condition for non-clan members as defined above holds true. However, the (level) monotonicity condition for clan members does not necessarily hold, because the remainder for the clan in larger subgames v_t might be less than the remainder for the clan in smaller subgames v_s.

Definition 12.38. *Let $v \in TMC^{N,m}_C$. A (level) payoff vector $b \in C(v)$ is* ***bi-limas extendable*** *if there exists a bi-limas $[a^t_{ij}]^{t\in\mathcal{M}^{N,1_C}}_{j\in N, j\in M^t_i}$ such that $b_{ij} = a^m_{ij}$ for each $i \in N$ and $j \in M^+_i$.*

Theorem 12.39. *Let $v \in TMC^{N,m}_C$ be such that inequality (12.23) holds for all $s, t \in \mathcal{M}^{N,1_C}$ such that $s \leq t$ and for each $i \in car(s_{N\setminus C})$. Then, there exist (level) payoff vectors $b \in C(v)$ which are extendable to a bi-limas.*

Proof. By Theorem 12.27 each (level) payoff vector $b \in C(v)$ is of the form

$$b_{ij} = \begin{cases} \alpha_{im_i}(v(m) - v(m_{-i}, m_i - 1)) & , \; i \in N \setminus C, \; j = m_i; \\[2ex] \alpha_{ij}(v(m_{-i}, j) - v(m_{-i}, j-1)) & , \; i \in N \setminus C, \\ \quad -\alpha_{i,j+1}(v(m_{-i}, j+1) - v(m_{-i}, j)) & \quad\; j \in M^+_i \setminus \{m_i\}; \\[2ex] \beta_i[v(m) - \sum\limits_{i \in N\setminus C} \alpha_{i1}(v(m) - v(m_{-i}, 0))] & , \; i \in C, \; j = 1, \end{cases}$$

where $\alpha_{ij} \in [0, 1]$ for each $i \in N \setminus C$ and $j \in M^+_i$, and $\beta \in \Delta(C)$, i.e. $\beta_i \geq 0$ for each $i \in C$ and $\sum\limits_{i\in C} \beta_i = 1$.

Consider the particular matrix $\alpha = (\alpha_{ij})_{i\in N\setminus C, j\in M^+_i}$ with $\alpha_{ij} = 0$ for each $i \in N \setminus C$ and each $j \in M^+_i \setminus \{1\}$. Denote $(\alpha_{i1})_{i\in N\setminus C}$ by $\tilde{\alpha}$.

Consider a core element $\tilde{b} \in C(v)$ corresponding to $\tilde{\alpha}$ and β. Note that $\tilde{b} = \psi^{\alpha,\beta}(v)$. Define for each $s \in \mathcal{M}^{N,1_C}$, each $i \in car(s_{N\setminus C})$ and each $j \in M_i^s$:

$$a_{ij}^s = \begin{cases} \alpha_{i1}(v(s) - v(s_{-i}, 0)) & , i \in car(s_{N\setminus C}),\ j = 1; \\ 0 & , i \in car(s_{N\setminus C}), \\ & \ \ j \in \{2, \ldots, s_i\}; \\ \beta_i[v(s) - \sum\limits_{k \in N\setminus C} \alpha_{k1}(v(s) - v(s_{-k}, 0))] & , i \in C,\ j = 1. \end{cases}$$

By Theorem 12.27, $[a_{ij}^s]_{i \in N\setminus C, j \in M_i^+}^{s \in \mathcal{M}^{N,1_C}}$ is a bi-limas. Now, note that $\tilde{b}_{ij} = a_{ij}^m$ for each $i \in N\setminus C$ and each $j \in M_i^+$. Hence, $\tilde{b}$ is bi-limas extendable.

Clearly, in case $m_{N\setminus C} = 1_{N\setminus C}$, a bi-limas coincides with a bi-mas and thus, the subclass of multi-choice total clan games considered in Theorem 12.36 coincides with the class of traditional total clan games. Hence, Theorem 12.39 says in this case that each core element of such a game is extendable to a bi-mas.

References

1. Arin, J. and V. Feltkamp (2002): Lorenz undominated allocations for TU-games: The weighted coalitional Lorenz solutions, Social Choice and Welfare 19, 869-884.
2. Arin, J. and E. Inarra (2001): Egalitarian solutions in the core, International Journal of Game Theory 30, 187-193.
3. Arin, J. and E. Inarra (2002): Egalitarian sets for TU-games, International Game Theory Review 4, 183-199.
4. Aubin, J.P. (1974): Coeur et valeur des jeux flous à paiements latéraux, Comptes Rendus de l'Académie des Sciences Paris 279 A, 891-894.
5. Aubin, J.P. (1981): Cooperative fuzzy games, Mathematics of Operations Research 6, 1-13.
6. Aubin, J.P. (1982): Mathematical Methods of Game and Economic Theory, North-Holland, Amsterdam.
7. Aumann, R. and L. Shapley (1974): Values of Non-atomic Games, Princeton University Press, Princeton.
8. Azrieli, Y. and E. Lehrer (2005): Extendable cooperative games, Journal of Public Economic Theory, forthcoming.
9. Azrieli, Y. and E. Lehrer (2006): On some families of cooperative fuzzy games, International Journal of Game Theory, forthcoming.
10. Azrieli, Y. and E. Lehrer (2007): Market games in large economies with a finite number of types, Economic Theory 31, 327-342.
11. Bennett, E. and M. Wooders (1979): Income distribution and firm formation, Journal of Comparative Economics 3, 304-317.
12. Bergantiños, G. and J. Massó (1996): Notes on a new compromise value: The χ-value, Optimization 38, 277-286.
13. Bhattacharya, A. (2004): On the equal division core, Social Choice and Welfare 22, 391-399.
14. Billot, A. (1995): Economic Theory of Fuzzy Equilibria, Springer-Verlag, Heidelberg.
15. Biswas, A., T. Parthasarathy, J. Potters, and M. Voorneveld (1999): Large cores and exactness, Games and Economic Behavior 28, 1-12.

16. Bondareva, O.N. (1963): Certain applications of the methods of linear programing to the theory of cooperative games, Problemy Kibernetiki 10, 119-139 (in Russian).
17. Branzei, R., D. Dimitrov, and S. Tijs (2003): Convex fuzzy games and participation monotonic allocation schemes, Fuzzy Sets and Systems 139, 267-281.
18. Branzei, R., D. Dimitrov, and S. Tijs (2004): Egalitarianism in convex fuzzy games, Mathematical Social Sciences 47, 313-325.
19. Branzei, R., D. Dimitrov, and S. Tijs (2004): Hypercubes and compromise values for cooperative fuzzy games, European Journal of Operational Research 155, 733-740.
20. Branzei, R., D. Dimitrov, and S. Tijs (2004): A new characterization of convex games, CentER Discussion Paper 2004-109, Tilburg University.
21. Branzei, R., D. Dimitrov, and S. Tijs (2006): The equal split-off set for cooperative games, Banach Center Publications 71, 39-46.
22. Branzei, R., D. Dimitrov, and S. Tijs (2006): Convex games versus clan games, CentER Discussion Paper 2006-58, Tilburg University.
23. Branzei, R., N. Llorca, J. Sánchez-Soriano, and S. Tijs (2007): Egalitarianism in multi-choice games, CentER Discussion Paper 2007-55, Tilburg University.
24. Branzei, R., N. Llorca, J. Sánchez-Soriano, and S. Tijs (2007): Multi-choice total clan games: characterizations and solution concepts, CentER Discussion Paper 2007-77, Tilburg University.
25. Branzei, R. and S. Tijs (2001): Additivity regions for solutions in cooperative game theory, Libertas Mathematica XXI, 155-167.
26. Branzei, R. and S. Tijs (2001): Cooperative games with a simplicial core, Balkan Journal of Geometry and its Applications 6, 7-15.
27. Branzei, R., S. Tijs, and J. Timmer (2001): Information collecting situations and bi-monotonic allocation schemes, Mathematical Methods of Operations Research 54, 303-313.
28. Branzei, R., S. Tijs, and J. Zarzuelo (2007): Convex multi-choice cooperative games and their monotonic allocation schemes, CentER Discussion Paper 2007-54, Tilburg University.
29. Brink, van den, R. (1994): A note on the τ-value and τ-related solution concepts, FEW Research Memorandum 652, Tilburg University, The Netherlands.
30. Brink, van den, R. (2002): A comment on the χ-value, Mimeo, Tilburg University, The Netherlands.
31. Butnariu, D. and E.P. Klement (1993): Triangular Norm-Based Measures and Games with Fuzzy Coalitions, Kluwer Academic Publishers, Dordrecht.
32. Calvo, E., E. Gutiérrez, and J.C. Santos (2000): The multichoice consistent value, International Journal of Game Theory 29, 177-188.
33. Calvo, E. and J.C. Santos (2000): A value for multichoice games, Mathematical Social Sciences 40, 341-354.

34. Calvo, E. and J.C. Santos (2001): A value for mixed action-set games, International Journal of Game Theory 30, 61-78.
35. Choquet, G. (1953/54): Theory of capacities, Annales de l'Institut Fourier 5, 131-295.
36. Curiel, I. (1997): Cooperative Game Theory and Applications: Cooperative Games Arising from Combinatorial Optimization Problems, Kluwer Academic Publishers, Dordrecht.
37. Curiel, I. and S. Tijs (1991): The minimarg and the maximarg operators, Journal of Optimization Theory and Applications 71, 277-287.
38. Derks, J. (1992): A short proof of the inclusion of the core in the Weber set, International Journal of Game Theory 21, 149–150.
39. Derks, J. and J. Kuipers (2002): On the number of extreme points of the core of a transferable utility game, in: P. Borm and H. Peters (Eds.), Chapters in Game Theory. In Honor of Stef Tijs, Kluwer Academic Publishers, pp. 83-97.
40. Derks, J. and H. Peters (1993): A Shapley value for games with restricted coalitions, International Journal of Game Theory 21, 351-360.
41. Dragan, I., J. Potters, and S. Tijs (1989): Superadditivity for solutions of coalitional games, Libertas Mathematica 9, 101-110.
42. Driessen, T. (1988): Cooperative Games, Solutions and Applications, Kluwer Academic Publishers, Dordrecht.
43. Driessen, T. and S. Tijs (1983): The τ-value, the nucleous and the core of a subclass of games, Methods of Operations Research 46, 395-406.
44. Driessen, T. and S. Tijs (1985): The τ-value, the core and semiconvex games, International Journal of Game Theory 14, 229-247.
45. Dutta, B. (1990): The egalitarian solution and reduced game properties in convex games, International Journal of Game Theory 19, 153-169.
46. Dutta, B. and D. Ray (1989): A concept of egalitarianism under participation constraints, Econometrica 57, 403-422.
47. Dutta, B. and D. Ray (1991): Constrained egalitarian allocations, Games and Economic Behavior 3, 403-422.
48. Ford, L.R. and D.R. Fulkerson (1956): Maximal flow through a network, Canadian Journal of Mathematics 8, 399-404.
49. Faigle, U. and W. Kern (1992): The Shapley value for cooperative games under precedence constraints, International Journal of Game Theory 21, 249-266.
50. Fukuda, E., S. Ishihara, S. Muto, S. Tijs, and R. Branzei (2005): Cooperative fuzzy games arising from economic situations, Fuzzy Economic Review 10, 3-15.
51. Gerard-Varet, L.A. and S. Zamir (1987): Remarks on the set of reasonable outcomes in a general coalition function form game, International Journal of Game Theory 16, 123-143.
52. Gillies, D.B. (1953): Some Theorems on n-person Games, Ph.D. Thesis, Princeton University Press, Princeton.

53. Grabisch, M. and F. Lange (2007): Games on lattices, multichoice games and the Shapley value: A new approach, Mathematical Methods of Operations Research 65, 153-167.
54. Grabisch, M. and L. Xie (2007): A new investigation about the core and the Weber set of multichoice games, Mathematics of Operations Research, forthcoming.
55. Harsanyi, J. (1959): A bargaining model for the cooperative n-person game, Annals of Mathematics Study 40, 325-355.
56. Hart, S. and A. Mas-Colell (1989): Potential, value and consistency, Econometrica 57, 589-614.
57. van Heumen, H. (1984): The μ-Value: A Solution Concept for Cooperative Games, Master Thesis, University of Nijmegen (in Dutch).
58. Hokari, T. (2000): Population monotonic solutions for convex games, International Journal of Game Theory 29, 327-338.
59. Hougaard, J., B. Peleg, and L. Thorlund-Petersen (2001): On the set of Lorenz-maximal imputations in the core of a balanced game, International Journal of Game Theory 30, 147-165.
60. Hsiao, C.-R. and TES Raghavan (1992): Monotonicity and dummy free property for multi-choice cooperative games, International Journal of Game Theory 21, 301-312.
61. Hsiao, C.-R. and TES Raghavan (1993): Shapley value for multi-choice cooperative games (I), Games and Economic Behavior 5, 240-256.
62. Ichiishi, T. (1981): Super-modularity: Applications to convex games and to the greedy algorithm for LP, Journal of Economic Theory 25, 283-286.
63. Kalai, E. and D. Samet (1987): On weighted Shapley values, International Journal of Game Theory 16, 205-222.
64. Kalai, E. and E. Zemel (1982): Totally balanced games and games of flow, Mathematics of Operations Research 7, 476-479.
65. Klijn, F., M. Slikker, and J. Zarzuelo (1999): Characterizations of a multi-choice value, International Journal of Game Theory 28, 521-532.
66. Lebret, A. and A. Ziad (2001): Fuzzy cooperative games and politikac models, GEMMA Working Paper 2001-1, Université de Caen Basse-Normandie.
67. Llerena, F. and Rafels, C. (2007): Convex decomposition of games and axiomatizations of the core and the D-core, International Journal of Game Theory 35, 603-615.
68. Lucas, W. (1968): A game with no solution, Bulletin of the American Mathematical Society 74, 237-239.
69. Luce, R. and H. Raiffa (1957): Games and Decisions, John Wiley & Sons.
70. Martínez-de-Albéniz, F.J. and C. Rafels (2002): A note on Shapley's convex measure games, Mimeo, University of Barcelona.
71. Martinez-Legaz, J.E. (2006): Some characterizations of convex games, in: A. Seeger (Ed.), Recent Advances in Optimization, Lecture Notes in

Economics and Mathematical Systems 563, Springer Verlag, Heidelberg, pp. 293-303.
72. Milnor, J.W. (1952): Reasonable outcomes for n-person games, Research Memorandum RM 916, The RAND Corporation, Santa Monica.
73. Molina, E. and J. Tejada (2002): The equalizer and the lexicographical solutions for cooperative fuzzy games: characterization and properties, Fuzzy Sets and Systems 125, 369-387.
74. Mondcrer, D., D. Samet, and L. Shapley (1992): Weighted Shapley values and the core, International Journal of Game Theory 21, 27–39.
75. Moulin, H. (1995): On additive methods to share joint costs, Japanese Economic Review 46, 303-332.
76. Muto, S., S. Ishihara, E. Fukuda, S. Tijs, and R. Branzei (2006): Generalized cores and stable sets for fuzzy games, International Game Theory Review 8, 95-109.
77. Muto, S., M. Nakayama, J. Potters, and S. Tijs (1988): On big boss games, Economic Studies Quarterly 39, 303-321.
78. von Neumann, J. and O. Morgenstern (1944): Theory of Games and Economic Behavior, Princeton University Press, Princeton.
79. Nisan, N., T. Roughgarden, and E. Tardos (2007): Algorithmic Game Theory, Cambridge University Press.
80. Nishizaki, I. and M. Sakawa (2001): Fuzzy and Multiobjective Games for Conflict Resolution, Physica-Verlag, Heigelberg.
81. Nouweland, van den, A. (1993): Games and Graphs in Economic Situations, PhD Thesis, Tilburg University.
82. Nouweland, van den, A., J. Potters, S. Tijs, and J. Zarzuelo (1995): Cores and related solution concepts for multi-choice games, Mathematical Methods of Operations Research 41, 289-311.
83. Owen, G. (1972): Multilinear extensions of games, Management Science 18, 64-79.
84. Owen G. (1995): Game Theory, 3rd ed., Academic Press, San Diego.
85. Peleg, B. (1986): On the reduced game property and its converse, International Journal of Game Theory 15, 187-200.
86. Peleg, B. and P. Sudhölter (2004): Introduction to the Theory of Cooperative Games, Springer-Verlag, Heidelberg.
87. Peleg, B. and P. Sudhölter (2007): Introduction to the Theory of Cooperative Games, 2nd ed., Springer-Verlag, Heidelberg.
88. Peters, H. and H. Zank (2005): The egalitarian solution for multichoice games, Annals of Operations Research 137, 399-409.
89. Potters, J., R. Poos, S. Tijs, and S. Muto (1989): Clan games, Games and Economic Behavior 1, 275-293.
90. Rafels, C. and S. Tijs (1997): On cores of cooperative games and the stability of the Weber set, International Journal of Game Theory 26, 491-499.
91. Rasmusen, E. (2007): An Introduction to Game Theory, 4th ed., Blackwell Publishing.

92. Rawls, J. (1971): A Theory of Justice, Harvard University Press.
93. Roth, A. (1988): The Shapley Value: Essays in Honor of Lloyd S. Shapley, Cambridge University Press, Cambridge.
94. Ruiz, L.M., F. Valenciano, and J.M. Zarzuelo (1996): The least square prenucleolus and the least square nucleolus. Two values for TU games based on the excess vector, International Journal of Game Theory 25, 113-134.
95. Sakawa, M. and I. Nishizaki (1994): A lexicographical concept in an n-person cooperative fuzzy game, Fuzzy Sets and Systems 61, 265-275.
96. Schmeidler, D. (1969): The nucleolus of a characteristic function game, SIAM Journal of Applied Mathematics 17, 1163-1179.
97. Schmeidler, D. (1972): Cores of exact games I, Journal of Mathematical Analysis and Applications 40, 214-225.
98. Selten, R. (1972): Equal share analysis of characteristic function experiments, in: Sauermann, H. (Ed.), Contributions to Experimentation in Economics, Vol. 3, J.C.B. Mohr (Paul Siebeck), Tübingen, pp. 130-165.
99. Sen, A.K. (1973: On Economic Inequality, Clarendon Press.
100. Seo, F., M. Sakawa, and I. Nishizaki (1995): Cooperative Fuzzy Games for International Conflict Solving, Theory and Decision Library, Series D, Kluwer Academic Publishers.
101. Shaked, M. and J.G. Shathikumar (1990): Parametric stochastic convexity and concavity of stochastic processes, Annals of the Institute of Statistical Mathematics 42, 509-531.
102. Shapley, L.S. (1953): A value for n-person games, Annals of Mathematics Studies 28, 307-317.
103. Shapley, L.S. (1967): On balanced sets and cores, Naval Research Logistics Quarterly 14, 453-460.
104. Shapley, L.S. (1971): Cores of convex games, International Journal of Game Theory 1, 11-26.
105. Shapley, L. and M. Shubik (1969): On market games, Journal of Economic Theory 1, 9-25.
106. Shapley, L. and M. Shubik (1975): Competitive Outcomes in the Cores of Market Games, International Journal of Game Theory 4, 229-237.
107. Sprumont, Y. (1990): Population monotonic allocation schemes for cooperative games with transferable utility, Games and Economic Behavior 2, 378-394.
108. Tijs, S. (1981): Bounds for the core and the τ-value, in: Moeschlin, O. and D. Pallaschke (Eds.), Game Theory and Mathematical Economics, North-Holland, Amsterdam, pp. 123-132.
109. Tijs, S. (1987): An axiomatization of the τ-value, Mathematical Social Sciences 13, 177-181.
110. Tijs, S. (2003): Introduction to Game Theory, Hindustan Book Agency, New Delhi.
111. Tijs, S. (2005): The first steps with Alexia, the average lexicographic value, CentER Discussion Paper 2005-123, Tilburg University.

112. Tijs, S. and R. Branzei (2002): Additive stable solutions on perfect cones of cooperative games, International Journal of Game Theory 31, 469-474.
113. Tijs, S. and R. Branzei (2004): Various characterizations of convex fuzzy games, TOP 12, 399–408.
114. Tijs, S., R. Branzei, S. Ishihara, and S. Muto (2004): On cores and stable sets for fuzzy games, Fuzzy Sets and Systems 146, 285-296.
115. Tijs, S., R. Branzei, S. Muto, S. Ishihara, and E. Fukuda (2004): Fuzzy clan games and bi-monotonic allocation rules, Fuzzy Sets and Systems 146, 271-284.
116. Tijs, S. and F.A.S Lipperts (1982): The hypercube and the core cover of n-person cooperative games, Cahiers du Centre d'Études de Researche Opérationelle 24, 27-37.
117. Tijs, S., A. Meca, and M.A. Lopez (2005): Benefit sharing in holding situations, European Journal of Operational Research 162, 251-269.
118. Tijs, S. and G.J. Otten (1993): Compromise values in cooperative game theory, TOP 1, 1-51.
119. Tsurumi, M., T. Tanino, and M. Inuiguchi (2000): The core and the related solution concepts in cooperative fuzzy games, Journal of Japan Society for Fuzzy Theory and Systems 12, 193-202.
120. Tsurumi, M., T. Tanino, and M. Inuiguchi (2001): A Shapley function on a class of cooperative fuzzy games, European Journal of Operational Research 129, 596-618.
121. Tsurumi, M., T. Tanino, and M. Inuiguchi (2004): Axiomatic characterizations of the Shapley function on a class of cooperative fuzzy games, Central European Journal of Operations Research 12, 47-57.
122. Voorneveld, M., S. Tijs, and S. Grahn (2002): Monotonic allocation schemes in clan games, Mathematical Methods of Operations Research 56, 439-449.
123. Webb, N.J. (2007): Game Theory. Decisions, Interaction and Evolution, Springer-Verlag, Heidelberg.
124. Weber, R. (1988): Probabilistic values for games, in: Roth, A.E. (Ed.), The Shapley Value: Essays in Honour of Lloyd S. Shapley, Cambridge University Press, Cambridge, pp. 101-119.
125. Weiß, C. (1988): Die Erweiterung nach Choquet und der Shapley-Wert auf Fuzzy-Spielen, Master Thesis, University of Bielefeld.
126. Weiß, C. (2003): Games with Fuzzy Coalitions: Concepts Based on the Choquet Extension, PhD Thesis, University of Bielefeld.
127. Yanovskaya, E. (2006): Consistency of the egalitarian split-off set for TU and NTU games, Discussion Paper, St. Petersburg Institute for Economics and Mathematics, Russian Academy of Sciences.
128. Young, H.P. (1985): Monotonic Solutions of Cooperative Games, International Journal of Game Theory 14, 65-72.

Index

Printed in the United States
116903LV00001B/169-174/P